GLOBAL CHANGE AND OUR COMMON FUTURE

PAPERS FROM A FORUM

Ruth S. DeFries and Thomas F. Malone,
editors

Committee on Global Change
National Research Council

National Academy Press
Washington, D.C., 1989

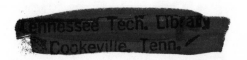

NOTICE: The project that is the subject of this report was approved by the Governing Board of the National Research Council, whose members are drawn from the councils of the National Academy of Sciences, the National Academy of Engineering, and the Institute of Medicine. The members of the committee responsible for the report were chosen for their special competences and with regard for appropriate balance.

This report has been reviewed by a group other than the authors according to procedures approved by a Report Review Committee consisting of members of the National Academy of Sciences, the National Academy of Engineering, and the Institute of Medicine.

The National Academy of Sciences is a private, nonprofit, self-perpetuating society of distinguished scholars engaged in scientific and engineering research, dedicated to the furtherance of science and technology and to their use for the general welfare. Upon the authority of the charter granted to it by the Congress in 1863, the Academy has a mandate that requires it to advise the federal government on scientific and technical matters. Dr. Frank Press is president of the National Academy of Sciences.

The National Academy of Engineering was established in 1964, under the charter of the National Academy of Sciences, as a parallel organization of outstanding engineers. It is autonomous in its administration and in the selection of its members, sharing with the National Academy of Sciences the responsibility for advising the federal government. The National Academy of Engineering also sponsors engineering programs aimed at meeting national needs, encourages education and research, and recognizes the superior achievements of engineers. Dr. Robert M. White is president of the National Academy of Engineering.

The Institute of Medicine was established in 1970 by the National Academy of Sciences to secure the services of eminent members of appropriate professions in the examination of policy matters pertaining to the health of the public. The Institute acts under the responsibility given to the National Academy of Sciences by its congressional charter to be an adviser to the federal government and, upon its own initiative, to identify issues of medical care, research, and education. Dr. Samuel O. Thier is president of the Institute of Medicine.

The National Research Council was organized by the National Academy of Sciences in 1916 to associate the broad community of science and technology with the Academy's purposes of furthering knowledge and advising the federal government. Functioning in accordance with general policies determined by the Academy, the Council has become the principal operating agency of both the National Academy of Sciences and the National Academy of Engineering in providing services to the government, the public, and the scientific and engineering communities. The Council is administered jointly by both Academies and the Institute of Medicine. Dr. Frank Press and Dr. Robert M. White are chairman and vice chairman, respectively, of the National Research Council.

Library of Congress Catalog Card No. 89-62950
International Standard Book Number 0-309-04089-2

Additional copies of this report are available from:

National Academy Press
2101 Constitution Avenue, NW
Washington, DC 20418

S015

FOREWORD

As the decade of the 1980s draws to a close, the world community of nations is on the brink of a new era. Our planet and the global environment are witnessing the most profound changes in the brief history of the human species. Human activity is the major agent of those changes-- depletion of stratospheric ozone, the threat of global warming, deforestation, acid deposition, the extinction of species, and others that have not yet become apparent.

The roots of global environmental change are embedded in the advances over the past few centuries in the understanding of the natural world and the utilization of natural resources. These scientific and technological advances have produced the driving forces of global change: an exponential growth in the world's population and even more rapid growth in the potential for humans to transform natural resources into goods and services to sustain that population. Over a period of half a million years, the human population has grown to a total of 5 billion individuals. Already, 40 percent of the planet's photosynthetic productivity is being used, diverted, or wasted. But already, too, there is a marked shortfall in meeting the basic human needs of more than a quarter of the world's population.

During the approximately 4000 days that remain before the dawn of the third millennium, Planet Earth will be asked to accommodate another billion people--approximately equivalent to the current populations of Africa, North America, and Europe combined. Within the next 50 years, we must somehow learn to feed, clothe, house, educate, and meaningfully employ an additional 5 billion individuals--the current population of the entire world. Over 90 percent of this increase will take place in developing countries.

To accommodate the doubling of the world's population at an acceptable standard of living, a 5- to 10-fold increase in the productive capacity of the world's agriculture and industry will be required. This is attainable, in principle, through scientific and technological progress, provided humankind makes a long series of small but correct decisions in the management of its affairs. It is theoretically possible for productive capacity to increase 2-fold in 1 decade in developing countries, and in 2 to 3 decades in developed countries. For this challenging possibility to become a reality, it is clear that humans must cast off their

unconscious role as the primary agent of global change and replace it
with a conscious role as prudent manager of change.

A special responsibility rests with the scientific and technological
community, which has developed the knowledge base that brings us to this
critical juncture. Now, the role of this community is to develop the
knowledge base upon which local, national, and global policy decisions
can be constructed with confidence. This function, however, cannot be
performed independently of the larger society of which science and
technology are a part. Thus the Forum on Global Change and Our Common
Future was organized to further the dialogue with the public on the key
issues related to describing, understanding, anticipating, and responding
to the dynamic interactions among the great interlocking physical,
chemical, biological, and social systems that regulate Planet Earth's
unique environment for life and determine the changes in the total earth
system.

A grand convergence of natural scientists, engineers, social scientists, and decision makers will be required worldwide. The World
Commission on Environment and Development has argued persuasively that
preservation and enhancement of the quality of the human environment,
wise stewardship of natural resources, and socioeconomic development are
inextricably related and mutually supportive. Moreover, the commission
has maintained that a guiding principle in successfully managing these
linked elements of global change is found in the concept of "sustainable
development"--development to meet the needs and aspirations of the
present generation without foreclosing options for future generations.
This concept implies balanced development between the developed and the
developing worlds to achieve intragenerational equity while protecting
the natural resource base to ensure intergenerational equity.

As revealed in the pages that follow, the tools and techniques for
understanding and responding to global environmental change are within
reach. It is not acceptable to defer action until all scientific issues
have been resolved unambiguously, nor is it advisable to undertake actions when the knowledge base is premature. There are salutary actions
that can be justified on narrower grounds that also address global environmental concerns, and there are actions that are prudent even in the
face of residual uncertainty. To discriminate among these options, a
new, dynamic, and creative interaction between the scientific and technological community responsible for developing the knowledge base and
the decision-making institutions in the public and private sectors is
needed. The participation of the social sciences must be strengthened;
the role of engineering needs to be made more explicit. Nationally and
internationally, institutional renewal, adaptation, and innovation will
be required in both the knowledge-developing and decision-making domains.
An unprecedented degree of cooperation within the world community of
nations will be needed.

Basic to confronting the challenge of global change is a fundamental
reorientation in the way of thinking among individuals everywhere.
Ultimately, it is the aggregate effect of individual actions that will
maintain Planet Earth's unique environment for life. Individuals shape
the collective consciousness; individual consumers make choices that

determine industrial policy; individual citizens form the political will to manage change.

We are rapidly approaching the end of the 5-century era within which the ebb and flow of military might and economic strength have been the key determinants of power, so perceptively described by Paul Kennedy in his book titled <u>The Rise and Fall of the Great Powers</u> (Random House, 1987). This era was brief--occupying only 1 percent of the span of years known as the period of Modern Man--a scant 0.1 percent of the tenure of <u>Homo sapiens</u> on Planet Earth. We need to adopt a time perspective that recognizes that just as Planet Earth is not more than halfway through its life expectancy, so also should <u>Homo sapiens</u> be viewed as not being more than halfway through its life expectancy. We are capable of managing our global affairs in a manner that looks forward to another half-million years of survival for our species. Surely, the human species is no less wise than were the dinosaurs.

As we near the third millennium, an increasingly interdependent world is at a critical watershed. Never before have we humans as a species, and as individual men and women, had such an opportunity to shape our common future. This forum is intended as a contribution to the wide-ranging discussions of the challenges and opportunities to determine that future.

Ruth S. DeFries, National Research Council
Thomas F. Malone, St. Joseph College and
Immediate Past President, Sigma Xi

PREFACE

Over the past few years, scientists, politicians, and the public have become increasingly aware that human activities are profoundly changing the global environment, with potentially severe consequences for human welfare. Almost daily, the media report on some aspect of global change --climate warming, deforestation, acid deposition, species extinction, depletion of stratospheric ozone, or other changes in the earth system. If the world's population continues to grow and if development proceeds according to current trends, we are told, the natural resource base on which our standard of living depends will be unavailable for future generations.

The National Research Council's Committee on Global Change recognizes that public understanding of the scientific issues of global change and the implications for policy is crucial for an informed, rational approach to addressing the complex issues of global change. Thus the committee embarked on organizing the Forum on Global Change and Our Common Future, held on May 2-3, 1989, at the National Theatre in Washington, D.C. The need for such a public forum was so widely recognized that the Smithsonian Institution, the American Association for the Advancement of Science, and Sigma Xi, the Scientific Research Society, joined the National Academy of Sciences in cosponsoring the event.

The objectives of the forum were threefold: (1) to present to the public a balanced and authoritative view of the wide range of global change issues, including the science of the earth system, the impacts of global change on society, and the implications for public policy; (2) to describe developments in the emerging interdisciplinary approach to the study of the earth system, aimed toward developing the knowledge base on which rational public policy decisions on global change can be pursued; and (3) to delineate the social, political, and economic framework within which the scientific and technological issues and the policy options need to be explored.

The forum grew out of several developments over the past decade. The international scientific community, in response to the alarming and overwhelming evidence that the earth system is changing in ways that are not fully understood, is embarking on an ambitious and long-term research program. The International Geosphere-Biosphere Program (IGBP), launched by the International Council of Scientific Unions (ICSU) in 1986, aims "to describe and understand the interactive physical, chemical, and

viii

biological processes that regulate the total earth system, the unique
environment it provides for life, the changes that are occurring in that
system, and the manner by which these changes are influenced by human
activities." The IGBP and other international programs and national
efforts, including the U.S. Global Change Research Program, collectively
constitute a new, interdisciplinary approach to the study of the earth
system, with the ultimate objective being to predict changes in the
system fundamental to human well-being.

Meanwhile, the United Nation's World Commission on Environment and
Development, under the leadership of Mme. Gro Harlem Brundtland, prime
minister of Norway, addressed the broad array of social, economic, and
political issues associated with "sustainable development"--development
to meet the needs and aspirations of the present generation without
foreclosing options for future generations. Their findings were published in 1987 in the notable book Our Common Future (Oxford University
Press).

Political interest in these issues of global environment and sustainable development has quickened all over the world. An intergovernmental panel--the Intergovernmental Panel on Climate Change--has been
convened within the United Nations to complement the nongovernmental ICSU
activities and to develop policy responses. In the United States, a
flurry of legislative proposals was introduced in the 100th Congress.
Throughout the world, these issues are being addressed at the highest
levels of government.

At the forum, the opening address by William Ruckelshaus, in which he
described society's stake in global change, set the stage for the two
days of discussion. The intrinsic variability of the global environment
over the geologic past was described by John Kutzbach as a prelude to an
examination of the earth system and its integrated components--the
atmosphere, the oceans, terrestrial ecosystems, and human interactions.
B. L. Turner II detailed the role of human activity in the global environment; his discussion was followed by explorations of several manifestations of global change: greenhouse warming, stratospheric ozone
depletion, deforestation, and acid deposition. The implications of
these consequences of human activity were developed by Paul Ehrlich.

The impacts of global change on human well-being, introduced by
Lester Brown, were explored on the morning of May 3 and included impacts
on agriculture and water resources and effects on biodiversity, sea
level, and industry. The afternoon was dedicated to a discussion of
public policy implications by an array of speakers from the international community and, in particular, from the Western Hemisphere. The
final evening was devoted to a panel summation, telecast to 52 Sigma Xi
chapters around the country.*

A highlight of the forum was the annual Benjamin Franklin Lecture by
Mme. Gro Harlem Brundtland. This challenging discussion, sponsored by

*For information about the availability of video recordings of the pre-
sentations by William Ruckelshaus, Mme. Gro Harlem Brundtland, and the
summary panel, contact Sigma Xi, 345 Whitney Avenue, New Haven, Connect-
icut 06511.

ix

the National Science Foundation, the National Academy of Sciences, and the American Association for the Advancement of Science, conveyed the conviction of the World Commission on Environment and Development that a more prosperous and more secure future is within reach.

This volume of papers includes 21 of the 38 presentations given at the forum, as well as the address by Senator Albert Gore, Jr., given the evening before the forum. The full range of issues covered in the forum is listed in Appendix A.

Many people came together to organize the forum. First and foremost, thanks are due to the speakers for their time and effort in preparing their thoughtful presentations. Harold Mooney, chairman of the Committee on Global Change (Appendix B), originally conceived of the forum and aided its development. From the Smithsonian Institution, Thomas Lovejoy, Robert Hoffmann, and Judith Gradwohl were indispensable in developing the program. Cheryl LaBerge and her staff in the Office of Conference Services provided impeccable logistical support. Thanks go to Pat Curlin and James Rowe of the American Association for the Advancement of Science for their useful input, and to Ed Poziomek and Peter Lykos of Sigma Xi for organizing the teleconferencing. Mary Keeney and Nan Smith from the National Science Foundation were key in organizing the Franklin Lecture. From the National Academy of Sciences, June Ewing was crucial in all of the organizational aspects of the forum, as was John Perry for his useful insights and comments. For editorial help in preparing the manuscripts for publication, acknowledgments go to Doris Bouadjemi and Susan Maurizi.

ACKNOWLEDGMENTS

The Forum on Global Change and Our Common Future was organized by the National Academy of Sciences; the Smithsonian Institution, in cooperation with the U.S. Committee for Man and the Biosphere; the American Association for the Advancement of Science; and Sigma Xi, the Scientific Research Society.

The annual Benjamin Franklin Lecture, held in 1989 in conjunction with the forum, is a featured activity of the National Science and Technology Week, an event created by the National Science Foundation to help educate the public and encourage America's future scientists and engineers. The Franklin Lecture is sponsored by the American Association for the Advancement of Science, the National Science Foundation, and the National Academy of Sciences.

Financial support for the forum was provided by the Business Roundtable, the Arthur L. Day Fund of the National Academy of Sciences, the Geraldine R. Dodge Foundation, the Department of Energy, the Environmental Protection Agency, the National Aeronautics and Space Administration, the National Oceanic and Atmospheric Administration, the National Science Foundation, the Tinker Foundation, and the U.S. Committee for Man and the Biosphere. The Rockefeller Foundation contributed to the support of the teleconferencing.

The views presented in this volume are not necessarily those of the organizing or sponsoring institutions.

CONTENTS

xi

PART A   SOCIETY'S STAKE IN GLOBAL CHANGE

TOWARD A GLOBAL ENVIRONMENTAL POLICY

William D. Ruckelshaus

It will come as no surprise when I say that politics is not entirely rational. It does not move in the world of crisp and precise analysis, but through more obscure channels. Politics dwells in symbol, in gesture, in metaphor. Some deplore this; I do not. Democracy is government by the people, and people are larger than their economics, or the numbers that describe them. They feel and they act on their feelings, and elected governments ignore feelings at their peril.

This is by way of prefacing my own feeling that as a metaphor for dealing with the current global environmental crisis, the word "management" leaves something to be desired. It is as if the environment were a horse that has suddenly become stubborn. We, of course, are the cowboy. This image puts us outside nature as its master, whereas the point of this crisis is surely that we are inside nature--are in fact both a contributor to the crisis and potentially its ultimate receptor.

The rhetoric of the environmental movement is partially to blame here. Seeking to convince the powerful to change their ways, many environmentalists have put forward an image of nature as vulnerable and helpless: the silent spring, the poor oil-soaked birds, the ravaged forests. Attractive animals and even particular ecosystems may be vulnerable, but Nature herself is not. Let us not forget that we are talking about a self-regulating system the size of a planet, 3 billion years old, about whose detailed workings we are still in profound ignorance.

The reason we are here in this hall, the reason that the scatter-brained attention of mankind has been focused, is because nature seems to be running a fever. We are the flu. Maybe that is a better metaphor, one that is more suitably humble. Our goal is not so much to manage planet earth as to make ourselves less like a pathogen and more like those helpful bacteria that dwell in our own guts. So make no mistake: It is not nature as a whole we are trying to protect; this is not about environmental protection. It may be about the survival of human society.

Science will figure very powerfully in how we do this, of course. Science is the necessary basis for political or social action. But the difficulty of converting scientific discovery into political action is a function both of the uncertainty of the science and the pain generated by the action. Given the current uncertainties as to the actual effects of the predicted rise in greenhouse gases, and the enormous social and technological effort that would be required to control them, it is fair to say that responding successfully to the global environmental crisis, and creating a fully sustainable world economy, will be a most difficult political enterprise, maybe <u>the</u> most difficult ever attempted.

Essentially, we would be trying to get a substantial proportion of the people of the world to change their behavior in order to possibly avert a set of changes that will mainly affect a world most of them will not live to see. One does not have to be an expert in politics to know that changes in human behavior do not ordinarily stem from such concerns.

Also, while models, such as the ones that now predict global warming, may convince scientists, who understand the models' assumptions and limitations, as a rule projections make poor politics. If you do not believe that, think of the clear and future danger of our national deficit. People will make enormous changes in their lives to escape a present danger, like war or a flood, or to improve their lot in an immediate way--by emigration, for example. But it is hard for people-- and hard even for the people who constitute governments--to change in response to something that might not happen for a long time, or might not happen at all.

Fortunately, we do have a response to such contingencies: We call it insurance. The analogy is apt. We think it prudent to pay insurance premiums so that if catastrophe strikes, we, or our survivors, will be better off than if there had been no insurance. Current resources foregone or spent to prevent the buildup of greenhouse gases are a sort of insurance premium.

And, as long as we are going to pay premiums, we might as well pay them in a fashion that will yield some dividend, in the form of greater efficiency, improved human health, or more widely distributed and sustainable prosperity. Such actions must include measures that will begin to reduce the rate of increase of carbon dioxide or acid rain or ozone by concentrating on those steps that most everyone will agree are reasonable under the circumstances. If we turn out to be wrong on greenhouse warming or ozone or acid rain, we still retain the dividend benefits. Also, no one complains to the insurance company when disaster does <u>not</u> strike.

That is the argument for some immediate, modest actions. On the other hand, something enormous may indeed be happening to our world. Our species may be pushing up against some immovable limits regarding combustion of fuels and ecosystem damage. Our usual tendency is to assume that if shortages or problems arise, we will discover a technological fix, or set of fixes, or that the normal workings of the market will adjust prices so as to solve the problem by product substitution. We may, for example, discover a cheap and nonpolluting source of energy.

It is comforting to imagine that we might get through this present crisis without much strain, to suppose, with Dickens' Mr. Micawber, that "something will turn up." Imagination is harmless; but <u>counting</u> on such a rescue may not be. We must at least consider the possibility that, besides those modest adjustments for the sake of prudence, we may have to prepare for far more dramatic changes. Doing this thinking now while we have the leisure to think is, in fact, another kind of insurance.

What would it take to move the world economy to true sustainability as recommended by the World Commission on Environment and Development? To answer that question we have to determine, first, what kind of change in consciousness would be required to maintain sustainability as a way of life. Such a change might include the adoption of the following benefits:

1. <u>The human species is part of nature. Its existence depends on its ability to draw sustenance from a finite natural world; its continuance depends on its ability to abstain from destroying the natural systems that regenerate this world</u>. This seems to be the major lesson of the environmental situation now as well as being a direct corollary of the second law of thermodynamics.

2. <u>Economic activity must account for all the environmental costs of production</u>. Environmental regulation has made a start here, but as yet a small one. The market has not been mobilized to preserve the environment at anywhere near its potential, with the result that an increasing amount of the "wealth" we think we create is in a sense stolen from our descendants.

3. <u>The maintenance of a livable world environment depends on the sustainable development of the entire human family</u>. This was one central finding of the World Commission on Environment and Development and appears to be the only reasonable option because of the well-documented impacts of population growth. Development stabilizes population; it is the only permanent solution we have discovered. If the four-fifths of humanity now in developing nations attempts to create wealth using the methods of the past, the result will at some point be unacceptable world ecological damage, such as accelerated ozone depletion or global warming.

If what is sustained is poverty, the result, given current population growth, will be mass death, social chaos, and accelerated environmental degradation of the type that results from poverty. Such situations also breed wars and the attendant danger that these will spread to the developed nations.

But changes in consciousness of this type do not come about simply because the arguments for them are good or because the alternatives are unpleasant. Neither will exhortation suffice. The central lesson of realistic policymaking is that most people and organizations change when it is in their interest to change, either because they derive some benefit from changing or because they incur sanctions when they do not, and the shorter the time span between the action and the benefit or sanction the better.

This is not mere cynicism. Although people will struggle and suffer for long periods to achieve a goal, it is unreasonable to expect most people to work against their immediate interests forever, especially in a democratic system, where their interests are so fundamental in guiding the government.

Changing interests requires three things. First, a clear set of <u>values</u> must be articulated by leaders in both the public and private sectors. Next, a set of <u>incentives</u> has to be established that will support those values. Finally, <u>institutions</u> must be developed that will effectively apply those motivators. The first is relatively easy, the second harder, the third hardest of all.

When we look at global environmental policy, we see that values similar to those described above are increasingly being articulated by political leaders throughout the world. In the past year, the president and the secretary of state of the United States, the premier of the Soviet Union, the prime minister of Britain, and the presidents of France and Brazil have all made major statements about global environmental problems. Most industrialized nations have a structure of national environmental law that reflects such values, and we now have a set of international conventions that does the same.

Yet mere acceptance of a set of values, while a necessary precursor, does not generate the necessary change in consciousness, nor does it change the environment. Although diplomats and lawyers may argue passionately over the form of words, talk is cheap. In the United States, for example, which has a set of environmental statutes second to none in their stringency, and where for the past 15 years, poll after poll has recorded the American people's desire for increased environmental protection, the majority of the population continues to participate in a most wasteful and polluting style of life.

The values are there; the appropriate incentives and the institutions are either absent or inadequate, and of course this is even more true of the earth as a whole.

The difficulties of moving from this situation stem from basic characteristics common to all the major industrial nations, the nations that must, because of their economic strength, their preeminence as polluters, and the share they claim of the world's resources, take the lead in any change of the present order. All of these nations are market system democracies, and it is apparent that an important part of the problem lies with something inherent in the free market economic system on the one hand, and with democracy on the other.

The economic problem is the familiar one of externalities, in which the environmental cost of producing a good or service is not accounted for in the price paid for it. As Kenneth Boulding has put it: "All of nature's systems are closed loops, while economic activities are linear and assume inexhaustible resources and 'sinks' in which to throw away our refuse." In willful ignorance, and in violation of the core principle of capitalism, we refuse to treat environmental resources as capital. We spend them as income and are as befuddled as any profligate heir when our checks start to bounce.

Closing the loops in economic systems--making people pay the full cost of the resource use--is the way to avoid this. That we have rarely done this in the industrialized world is related to the second problem, the problem of action in a democracy. Modifying the market to reflect environmental costs is largely a function of government. Those adversely affected by such modifications, although they may be a tiny minority of the population, often have a disproportionate influence on public policy. In general, the minority much injured will prove more formidable a lobbyist than the majority slightly benefited.

The interest problem is naturally exacerbated when dealing with pollution on a global scale. Elected representatives are even less likely to support short-term adverse effects on their constituencies when the immediate beneficiaries are residents of other lands. This reluctance is magnified even more by scientific uncertainty regarding the timing, origin, or importance of those benefits.

The question then, is whether the industrial democracies will be able to overcome the political constraints on bending the market system toward long-term sustainability. History suggests some answers, for there are a number of examples in which nations have been able to harmonize a variety of short-term interests with a longer-term goal.

War is, regrettably, the obvious example. A conflict, like the Second World War, that mobilizes the entire population, changes work patterns, manipulates and controls the prices and supply of standard goods, and reorganizes the nation's industrial plant demonstrates that things considered politically or economically impossible can be accomplished in a remarkably short time, given the belief that national survival is at stake.

Another example is found in the Marshall Plan for reconstructing Europe after World War II. In 1947 the United States spent nearly 3 percent of its gross domestic product (GDP) on this huge set of projects. Although the impetus for the plan came from fear of the expansion of Soviet influence into Western Europe, it established a precedent for massive investment in increasing the prosperity of foreign nations.

Besides these, there are numerous examples where belief systems many generations old changed rapidly under the press of necessity. These include the abandonment of feudalism by Japan and of slavery by the industrialized nations in the nineteenth century, and the retreat of imperialism and the development of the European Community in the twentieth century. In each of these, important interests were made to give way before national goals.

We should also not forget that of all the political and economic systems that have been devised, liberal democracies based on free enterprise appear to be the most capable of change. At any rate they appear to have survived the passing of all the others, and they now dominate the world.

If it is possible to change, how do we begin? Obviously government policy must lead the way, since market prices of commodities typically do not reflect the environmental costs of extracting and replacing them, nor do prices of energy from fossil fuels reflect the risks of climate change. And policy matters. In the case of global warming, for example, policies implemented soon and continued over the next decade could significantly affect the rate and extent of the greenhouse effect.

Policy must focus on changing incentives and perfecting institutions. If we do that, the values of sustainability will thrive and survive. If we do not, they will degrade along with the environment. The leaders in making these policy changes must be the developed nations, and they must begin with their domestic economies, which currently use the bulk of the world's resources.

If they do not they will have no credibility with the leaders of the developing world, a necessary prerequisite to achieving sustainable development. And that, of course, remains our greatest challenge.

Aid is both an answer and a perpetual problem. The total of official development assistance from the developed to the developing world stands at around $35 billion per year. This is not a great deal of money when one considers that if the United States now spent in foreign aid the same proportion of GNP it spent during the peak Marshall Plan years, the annual U.S. foreign aid expenditure would be $127 billion. For comparison, the United States spent $45 billion protecting shipping in the Persian Gulf.

There is no point, of course, in even thinking about the adequacy of aid to the undeveloped nations until the debt issue is resolved. The World Bank reported in 1988 that the 17 most indebted countries paid the rich nations and multilateral agencies $31.1 billion more than they received in aid. This obviously cannot go on, and debt-for-nature swapping alone will not solve the problem.

In most nations, we now realize, a prosperous rural society based on sustainable agriculture must be the prelude to any future development. To obtain that, land tenure reform will have to be instituted in many countries and basic international trading relationships will have to be redesigned to eliminate the ill effects on the undeveloped world of agricultural subsidies and tariff barriers in the rich nations.

This is another way of saying we must focus on what motivates people to live in an environmentally responsible manner. People will not grow crops when governments subsidize urban populations by keeping prices to farmers low. People will not stop having too many children if the labor of children is the only economic asset they have. People will not improve the land if they do not own it.

Negative sanctions against abusing the environment are similarly missing throughout much of the undeveloped world. In the short term, substantial amounts of aid could be focused directly on the environmental protection ministries of developing nations. These ministries are typically impoverished and ineffective, particularly in comparison to their countries' economic development and military institutions. To cite one example: The game wardens of Tanzania receive an annual salary equivalent to the price paid to poachers for two elephant tusks, one reason why that nation has lost two-thirds of its elephant population to the ivory trade in the last decade.

Finally, we must create and maintain institutions that will support the values and motivators that favor a sustainable world economy. This is a difficult task, for institutions are powerful in that they support some powerful interests, which usually includes supporting the status quo. On the other hand, free societies are good at creating effective institutions, and the transfer of power among their institutions, according to perceived social needs, is a fact of life.

The important international institutions in today's world are those concerned with money, with trade, and with national defense. Those who may despair of environmental concerns ever reaching this level of seriousness should recall that current institutions like NATO, the World Bank, and multinational corporations have fairly short histories. They were formed out of pressing concerns about continuing the expansion of wealth and maintaining national sovereignty. If concern for the environment becomes pressing on a comparative scale, comparative institutions will be developed.

To further this goal, three things are wanted. The first is money. The report of the World Commission on Environment and Development says: "The U.N. can and should be a source of significant leadership in the transition to sustainable development and in support of developing countries in effecting this transition." The annual budget of the United Nations Environment Programme is $30 million, a laughable amount considering its responsibilities.

If we are serious about sustainability, we will provide our central international environmental organization with serious money, preferably money derived from an independent source to reduce its political vulnerability. An international tax on certain uses of common world resources has been suggested as a means to this end.

The second thing is information. We require strong international institutions to collect, analyze, and report on environmental trends and risks. We need a global institution capable of answering questions of global importance.

The third thing is integration of effort. We obviously do not wish to create a monolithic bureaucracy, but neither can we afford redundancy and conflict in our efforts to solve common problems. On the aid front, this may become tragically absurd: Africa alone is currently served by 82 international donors and over 1700 private organizations. In 1980, in the tiny African nation of Burkina Faso (population 8 million) there were 340 independent aid projects under way. We need to form and strengthen coordinating institutions that combine the separate strengths of nongovernmental organizations, international bodies, and industrial groups and focus their efforts on global warming and on the short list of environmental priority issues identified by the World Commission on Environment and Development.

Finally, in creating the consciousness of advanced sustainability, we will have to redefine our concepts of political and economic feasibility. These are, after all, human constructs. They were different in the past; they will surely change in the future. But the earth is real, and we are obliged by the fact of our utter dependence on it to listen more closely than we have to its messages.

2

GLOBAL CHANGE AND OUR COMMON FUTURE
THE BENJAMIN FRANKLIN LECTURE*

Gro Harlem Brundtland

We are living in an historic transitional period in which awareness
of the conflict between human activities and environmental constraints is
literally exploding. This finite world will have to provide food and
energy and meet the needs of a doubled world population sometime in the
next century. It may have to sustain a world economy that is 5 to 10
times larger than the present one. It is quite clear that this cannot be
done by perpetuating present patterns.

In the never-ending human search for an improved habitat, for new
materials, new energy forms, and new processes, the constraints imposed
by depletion of natural resources and the pollution caused by the conversion of resources have brought mankind to a crossroads.

In spite of all the technological and scientific triumphs of the
present century, there have never been so many poor, illiterate, or
unemployed people in the world, and their numbers are growing. Close to
1 billion people are living in poverty and squalor, a situation that
leaves little choice in a struggle for life that often undermines the
conditions for life itself, the environment, and the natural resource
base.

We continue to live in a world where abundance exists side by side
with extreme need, where waste overshadows want, and where our very
existence is in danger due to mismanagement and over-exploitation of the
environment.

The undermining of respect for international obligations was one of
the many negative trends in international politics during the 1970s and
the early 1980s.

I believe that the threats to the global environment have the
potential to open our eyes and to make us accept that North and South
will have to forge an equal partnership. The threats to the global
climate prove beyond doubt that, if all do as they please in the short
run, we will all be losers in the long run. We need to develop a more
global mentality in charting the course toward the future, and we need
sound scientific advice and firm political and institutional leadership.

---

*This lecture was published in a slightly different form in Environment
31(5):16-20, 40-43 (June 1989) by Heldref Publications, 4000 Albemarle
St., N.W., Washington, D.C. 20016.

We face a grim catalog of environmental deterioration. We know that forests are vanishing. Every year 150,000 km^2 disappear. We are becoming increasingly aware of the spread of desert land. The yearly rate is 60,000 km^2. Good soil is being washed away or eroding at alarming rates. It is estimated that about 150 plant and animal species are becoming extinct every day, most of them unknown to laymen and specialists alike. The stratospheric ozone shield is in danger. And above and beyond all these signs of environmental crisis, the climate itself is threatened.

As the challenging dynamics of global change gradually become clearer, the role of the men and women of science in shaping our common future becomes more central. The interplay between the scientific process and the making of public policy is not a new phenomenon. Indeed, it has been a characteristic of most of the great turning points in human history. One need look back no further than the dawning of the nuclear age to conclude that names such as Enrico Fermi, Niels Bohr, Julius Oppenheimer, and Andrei Sakharov have influenced today's world just as much as Franklin D. Roosevelt, Joseph Stalin, Winston Churchill, Mahatma Gandhi, and Dag Hammarskjold.

It may be more important now than ever before in history for scientists to keep the doors of their laboratories open to political, economic, social, and ideological currents. The role of the scientist as an isolated explorer of the uncharted world of tomorrow must be reconciled with his role as a committed, responsible citizen of the unsettled world of the present.

The interaction between politics and science has been decisive in the pursuance of international consensus on the problem of stratospheric ozone depletion. The protocol hammered out in Montreal in September 1987, which provides for reducing chlorofluorocarbon (CFC) emissions by 50 percent over the next decade, could never have been achieved without a delicate balance between the most up-to-date scientific information, reliable industrial expertise, and committed political leadership against a background of strong and informed public interest.

The fact that new scientific data on the threat to the ozone layer have already prompted us to move beyond the 1987 accords only underlines my point: The scientist's chair is now firmly drawn up to the negotiating table, right next to that of the politician, the corporate manager, the lawyer, the economist, and the civic leader. Indeed, moving beyond compartmentalization and outmoded patterns to draw on the very best of our intellectual and moral resources from every field of endeavor lies at the very heart of the concept of sustainable development.

It is a rare privilege to discuss the challenges before us as we approach the end of a century that has brought more changes than the entire previous history of mankind. I do so emphasizing that U.S. leadership will be decisive if we are to succeed, on a global scale, in making the necessary changes. I do so with the greatest respect and admiration for the human and material resources of the United States, resources that can and must be mobilized for sustainable development if we are to overcome the interlocked environment and development crisis.

The United States has perpetually fostered human genius. Benjamin Franklin himself was a paragon of intellectual curiosity and versatility.

His inquisitive, insatiable mind was constantly on the lookout for knowledge and would have found it in a desert. His own words about learning are illustrative, and I quote: "In persons of a contemplative disposition, the most different things provoke the exercise of the imagination, and the satisfactions which often arise to them thereby are a certain relief to the labor of the mind as well as to that of the body."

Had Franklin been alive today, he might have found a solution to the energy problem. He was actually very involved with the problem of energy efficiency. Franklin was the first scientist to study the Gulf Stream. He found that a vessel sailing from Europe to America could shorten the voyage by avoiding the Gulf Stream and that a thermometer could be used to determine the edge of it.

Today, the international agenda has grown more varied and complex, but also more promising. Advances are being made in a number of fields, including the easing of tensions between East and West with the ensuing gains for peace and security and the settlement of regional conflicts.

Should we not take advantage of this favorable climate and direct our efforts toward the critical environment and development issues facing us? Many of these problems cannot be solved within the confines of the nation state, nor by maintaining the dichotomy between friend and foe. We must increase communication and exchange and cultivate greater pluralism and openness.

In 1987, the World Commission on Environment and Development presented its report Our Common Future. The commission sounded an urgent warning: Present trends cannot continue. They must be reversed.

The World Commission did not, however, add its voice to that of those who are predicting continuous negative trends and decline. The commission's message is a positive vision of the future. Never before in our history have we had so much knowledge, technology, and resources. Never before have we had such great capacities. The time and the opportunity have come to break out of the negative trends of the past.

What we need are new concepts and new values based on a new global ethic. We must mobilize political will and human ingenuity. We need closer multilateral cooperation based on the recognition of the growing interdependence of nations.

The World Commission offered the concept of sustainable development. It is a concept that can mobilize broader political consensus, one on which the international community can and should build. It is a broad concept of social and economic progress. The commission defined sustainable development as meeting the needs and aspirations of the present and future generations without compromising the ability of future generations to meet their needs. It requires political reform, access to knowledge and resources, and a more just and equitable distribution of wealth within and between nations.

Over the past couple of years some progress has been made in the environmental field, both in terms of raising consciousness and in terms of taking on particular challenges, such as in the Montreal Protocol on the ozone layer and the Basel Convention on hazardous wastes. However, the picture is very uneven, and the achievements far from justify complacency.

As far as development is concerned, however, the 1980s have been a lost decade. Although some countries have done well, there has been widespread economic retrogression in the Third World. Living standards have declined by one-fifth in sub-Saharan Africa since 1970. Unsustainable, crushing burdens of debt and reverse financial flows, depressed commodity prices, protectionism, and abnormally high interest rates have all created an extremely unfavorable international climate for development in the Third World.

Politically, economically, and morally, it is unacceptable that there should be a net transfer of resources from the poor countries to the rich. Paradoxically, the fact of the matter is that while close to 1 billion people are already living in poverty and squalor, the per capita income of some 50 developing countries has continued to decline over the past few years.

These trends will have to be reversed. As pointed out by the World Commission on Environment and Development, only growth can eliminate poverty. Only growth can create the capacity to solve environmental problems. But growth cannot be based on over-exploitation of the resources of developing countries. Growth must be managed to enhance the resource base on which these countries all depend. We must create external conditions that will help rather than hinder developing countries in realizing their full potential.

What we now need is <u>global consensus for economic growth in the 1990s</u>. It must comprise the following:

o Economic policy coordination that will promote vigorous noninflationary economic growth. Major challenges include reducing payment imbalances between the United States, Japan, and the Federal Republic of Germany and making the surpluses of Japan, the Federal Republic of Germany, and other countries increasingly available to developing countries.

From a world development point of view, the financial surpluses of the Organization for Economic Cooperation and Development (OECD) countries should increasingly be used for investments in developing countries rather than for financing private consumption in the major industrialized countries.

o Policies that will secure more stable exchange rates and increased access to markets on a global basis. Protectionism is a confrontational issue and a no-benefit game. Every year, protectionism costs the developing countries twice the total amount of development assistance they receive. The benefits of free trade both for the North and for the South are obvious.

o Policies that will sustain and improve commodity prices.

o Policies that encourage and support diversification of the economies of the developing countries. We need adjustment programs that are realistic. Their pace and sequence must be carefully tailored to the characteristics and the development priorities of the individual countries through a policy of dialog. More must be done to incorporate poverty concerns and environmental considerations into adjustment programs.

o Major new efforts that will reduce debt based on the recent Brady initiative. For debt owed to multilateral institutions, the scheme based on a Nordic proposal to soften interest payments on such loans has been taken up by the World Bank. We believe this and similar schemes should be extended in the future.

A very civilized, ancient legal provision on debt reads as follows: "If a man owes a debt, and the storm inundates his field and carries away the produce, or if the grain has not grown in the field, in that year he shall not make any return to the creditor, he shall alter his contract and he shall not pay interest for that year." This quote is taken from the Code of Hammurabi, King of Babylon, which dates from the year 2250 B.C. Four thousand years later the debt burdens, the environmental crisis, and the decline in the flows of resource transfers are trends that call for equally civilized considerations.

o In addition to our debt efforts, what is called for is increased development assistance, nothing short of a "Marshall Plan" for the poorer nations of the developing world, notably for Africa. I see no reason to conceal that while Norway has given around 1.1 percent of its GNP in official development assistance to developing countries in recent years, we are disappointed that the OECD average has declined to a meager 0.34 percent. Those donor countries that have been lagging behind in their official development assistance transfers should now make renewed efforts in line with their abilities.

The Soviet Union and Eastern Europe must also contribute to a far greater extent than they have done so far. The developing countries have been declaring their readiness to do their part in terms of policy reforms and constructive negotiations.

A global consensus for economic growth in the 1990s must be consistent with sustainable development. It must observe ecological constraints. There are no sanctuaries on this planet. If the next decade is to be truly a decade of response to the serious problems that confront the world, the issue of sustainable global development must receive special, and urgent, attention.

It is time for a global economic summit to launch a new era of international cooperation. Issues like the debt crisis, trade matters, resources for the international financial institutions, harnessing technology for global benefit, strengthening the United Nations system, and specific major threats to the environment, such as global warming, are becoming increasingly interrelated. Would it not be appropriate to consider both our economic and our environmental concerns together at such a summit, given the critical links between the two?

The Third World seems convinced that international poverty is not a mere aberration of international economic relations that can be corrected by minor adjustments, but rather the unspoken premise of the present economic order. Developing countries have had to produce more and sell more in order to earn more to service debt and finance imports. And the amount of coffee, cotton, or copper they have had to produce to buy a water pump, antibiotics, or a lorry has kept increasing.

This has led to over-taxation of the environment. It has fueled soil erosion and accelerated the cancerous processes of desertification and deforestation, which in turn have begun to threaten the genetic diversity that is the basis for tomorrow's biotechnology, agriculture, and food supply.

Biotechnology is a case in point. The effects of modern biotechnology on agriculture and food security in the Third World must be given special care and attention. Clearly, the production of enough food to feed a doubled world population is inconceivable without biotechnology. But there are inherent dangers that could, unless they are avoided, further widen the gap between poor and rich.

The benefits of plant breeding and plant varieties with greater resistance and more rapid growth potential have been and will continue to be immense. But these benefits may become available only to the rich, while the genes employed in the process often originate in developing countries that derive very little benefit from their use.

Strong international corporations may dominate this field. Legal protection and very firm rules regarding rights of ownership may reduce the availability of products that are important for nutrition and the prevention of famine.

Small-scale farmers in the Third World risk being victims in this process. Biotechnology may produce substitutes for their crops. They may lose income and the ability to provide for their families.

The industrialized countries have a responsibility for controlling market forces in this field and for promoting a more equitable sharing between developed and developing countries. The protection of intellectual property rights and royalties must be in a form that promotes research, provides for an equitable sharing of financial benefits between inventors and the country of genetic origin, and not least, makes the products of biotechnology available to those who need them.

We need to foster a stronger sense of collective responsibility and to make the international bodies we have created more effective. The time has come to seek more innovative structures for cooperation than those we have available at present. Stronger mandates for making binding decisions should be worked out.

The threats of global heating and climatic change may be the most severe threat to future development. Life on earth depends on the climate. Human settlement, food production, and industrial patterns are at stake.

The effects of climate change may be enormous. The impact may be greater and more drastic than any other challenge previously facing mankind, with the possible exception of the threat of nuclear war.

There is one big, decisive difference here. Whereas nuclear war can be avoided--and at present it seems more remote than at any time since World War II--we will be caught in the heat trap of global warming unless we reduce our consumption of fossil fuels.

We may be about to alter the entire ecological balance of the earth. The time span needed for plants and animals to adjust to a new climate is normally hundreds of years. However, unless drastic changes are made, the ecosystems will not be able to adjust. Deserts will spread. Crops will be lost. Last year's drought may not have been the result of climatic change, but what will happen if we experience 2 such dry summers, or 10 such dry summers, in succession? What will happen to food production? Can we conceive of a doubling of food prices, or even a scarcity of food in the industrialized countries? The developed countries may be able to cope in the short run as long as they can pay for necessary imports. But that option will soon be lost to the developing countries.

Can we conceive of the effects on low-lying countries if the sea level should rise according to predictions? Can we see any solutions to the political instability that will accompany increased migration as the number of environmental refugees continues to multiply?

All this may not happen, or it may not be that drastic. But the potential risks are so high that we cannot sit back hoping that the problems will solve themselves.

The present generation has a great responsibility. It is this generation that will have to set limitations on our own use of limited resources, in particular on the burning of fossil fuel. We must recognize that the earth's atmosphere is a closed system. We are not getting rid of our emissions. In fact the current system is like a car that pours out its gases into the driver's compartment.

We must tackle the myth that energy consumption must be allowed to grow unchecked. The industrialized countries have the greatest resources, both financially and technologically, to change production and consumption patterns. The developing countries will need much more energy in the future. Many of them have contributed only marginally to the greenhouse effect, and many of them will be most severely victimized by global heating. They must be allowed more time for adaption and a chance to increase their consumption.

We need concerted international action. There are certain imperatives that must be pursued with vigor as matters of the utmost urgency:

o We need to agree on regional strategies for stabilizing and reducing emissions of greenhouse gases. Reforestation efforts must be included as a vital part of the carbon equation.

o We must strongly intensify our efforts to develop renewable forms of energy. Renewable energy should become the foundation of the global energy structure during the twenty-first century.

It is quite clear that developing countries will need assistance to avoid making the same mistakes we have made. It is essential that energy-efficient technology be made available to developing countries when they cannot always pay market prices without assistance.

o We should urgently speed up our efforts on international agreements to protect the atmosphere. There are different views on how to proceed on this issue. I urge that negotiations to limit emissions be started immediately.

On March 11, 1989, 22 heads of state and government signed a declaration in the Hague that set a standard for future achievements to protect the atmosphere. In the Declaration of The Hague we called for more effective decisionmaking and enforcement mechanisms in international cooperation as well as greater solidarity among nations and between generations. The principles we endorsed were radical, but any approach that is less ambitious would not serve us.

The declaration calls for new international authority with real powers. On occasion the power must be exercised even if unanimity cannot be reached.

We must have defined standards and ensure compliance. We must have effective regulatory and supportive measures and uphold the rule of law.

Sharing the burden is essential. That is why we called for fair and equitable assistance to compensate those developing countries that will be most severely affected by a changing climate but that have contributed only marginally to global heating.

The Norwegian government very recently adopted a white paper on the follow-up of the World Commission's report. It has involved all ministries, not only that of the environment; it has implied change in attitudes and policies, and tough challenges for the heavy sectoral ministries--such as energy, industry, transportation, finance, foreign affairs, and trade--and the prime minister's office has been directly engaged in charting a cross-sectoral course for the future.

The issue of atmospheric pollution and climate change proved to be a very difficult one. It is difficult because Norway has been fortunate to have vast hydropower resources. We do not burn coal or oil to produce electricity. Any reduction of carbon dioxide (CO_2) emissions in Norway would involve transportation.

Many ask why Norway could make a difference when we cause only 0.2 percent of global CO_2 emissions. Should we impose limitations on ourselves even if other countries have not yet done so?

The Norwegian government has chosen to set out clear goals. I believe we are the first country to make a political commitment for reductions of CO_2 emissions.

o Norway sets a policy for stabilizing its emissions of CO_2 in the course of the 1990s and at the latest by the year 2000.

o The government presupposes that thereafter a reduction will be possible.

o Together with our reductions of CFCs and nitrogen oxides (NO_x), Norway will be able to reduce its total emissions of greenhouse gases by the turn of the century.

Clearly, the larger ecological issues--the ozone layer, global warming, and the sustainable utilization of the tropical forests--are tasks facing mankind as a whole. To finance these tasks we will need additional resources.

In the white paper, our major policy document on sustainable development, the Norwegian government is proposing, as a starting point, that industrialized countries allocate 0.1 percent of their gross domestic product to an International Fund for the Atmosphere. Such a fund should be created to help finance transitory measures in developing countries and reforestation projects. Ideally, all countries should take part in this. Everyone would then make their contribution.

Much work is needed to make this proposal operational, and it will be met with considerable reluctance. But unless we establish a set of international support mechanisms, there are fewer chances that we will be able to make the transition in time.

I have presented to you the essence of <u>Our Common Future</u>. To transform it into reality will require broad participation. Every single individual can make a difference. Changes are the sum of individual actions based on common goals.

A particular challenge goes to youth. More than ever before, we need a new generation--today's young people, who with new energy and dedication can turn ideas into reality.

Many of today's decisionmakers have yet to realize the peril in which this earth has been placed. I believe that <u>Our Common Future</u> can be an effective lever in the hands of youth and that it can transcend nationality, culture, ideology, and race. Youth will hold their governments responsible and accountable, and youth will be stalwarts for the foundation of their own future.

In closing, let me stress the need for all of us to view environmental problems in interdisciplinary terms, not in the narrow terms of specialization. The world is replete with projects that made excellent engineering sense but were economically disastrous or that were economically sound but environmentally catastrophic. The global environment cannot be separated from political, economic, and moral issues. Environmental concerns must permeate all decisions, from consumer choices through national budgets to international agreements. We must learn to accept the fact that environmental considerations are part of a unified management of our planet. This is our ethical challenge, our practical challenge, a challenge we all must take.

3

GLOBAL CHANGE AND CARRYING CAPACITY:
IMPLICATIONS FOR LIFE ON EARTH

Paul R. Ehrlich, Gretchen C. Daily, Anne H. Ehrlich,
Pamela Matson, and Peter Vitousek

Determining the long-term carrying capacity of Earth--that is, the number of people that the planet can support without irreversibly reducing its ability to support people in the future--is an exceedingly complex problem. About all we can be sure of now is that, with present and foreseeable technologies, the human population has already exceeded that capacity. Even today's 5.2 billion people can only be supported by a continuing depletion of humanity's one-time inheritance from the planet: nonrenewable resources including deep, rich agricultural soils, "fossil" groundwater, and the diversity of nonhuman species.

Carrying capacity is a function of characteristics of both the human species and the planet. Through cultural evolution, human beings may quickly shift their demand for and ability to extract different resources. At the same time, natural and anthropogenic processes change the distribution and abundance of resources in the short and medium term. This paper addresses the latter aspect of carrying capacity: the influence of global change on the planet's capability to support people over the next 20 to 100 years.

Carrying capacity can be broken down into a number of interacting elements, including food, energy, ecosystem services (such as provision of fresh water, flood control, and recycling of nutrients; Ehrlich and Ehrlich, 1981), the epidemiological environment, social structure, politics, and culture. Applying the reasoning of Liebig's "law of the minimum," overall carrying capacity is determined by whichever component yields the lowest carrying capacity. Much of our discussion here focuses on food because, although it may not ultimately be the limiting resource of human population size, food production is a crucial factor that is very sensitive to global change. In addition, basic human nutritional requirements are relatively inflexible and easy to quantify in contrast to other elements: there is no substitute for food.

It is especially critical to evaluate carrying capacity now because the human population has clearly exceeded local and regional carrying capacities in many parts of the world (FAO, UNFPA, and IIASA, 1982), as shown by an increasing failure of food production to keep pace with population growth. For the first time ever, moreover, carrying capacity has been exceeded globally. Furthermore, human population pressure is reducing carrying capacity directly through the unsustainable use and consequent destruction of natural habitat and agricultural land (Brown, 1988; Ehrlich and Ehrlich, 1988).

The human population has indirect impacts on carrying capacity as well. The magnitude of these impacts can be evaluated as the product of three interacting, multiplicative factors, of which population size is one. The other two factors are per capita consumption of resources (a measure of affluence) and some measure of the environmental damage generated by technologies used to provide each unit of consumption (Holdren and Ehrlich, 1974). Indirect impacts are causing global environmental changes that themselves influence the number of people the Earth can support. Of these changes, the greatest potential consequences for carrying capacity appear to reside in anthropogenic changes in the global climatic system.

Population growth thus contributes to a widening gap between the quantity of resources, especially food, needed by the human population and the amount that can be extracted from the planet. In the following, we discuss the reduction in carrying capacity that can be expected to result from direct human impacts on resources and the environment and from our indirect impacts on the climatic system.

DIRECT HUMAN IMPACTS

The Stanford Carrying Capacity Project has estimated that the human population now uses directly, coopts, or has destroyed approximately 40 percent of global net primary productivity on land, the basic food supply of all terrestrial animals (Vitousek et al., 1986). Humanity is not only exercising increasing control over this global food supply but is also undermining the capacity of photosynthesizing organisms to produce it.

The direct human impact on carrying capacity is especially evident on marginal land at both extremes of the moisture gradient. Arid and semiarid regions, particularly in Africa, are suffering severe degradation through desertification. A total of 27 million hectares of land-- an area the size of the state of Colorado--completely lose economic utility each year because of excessive human impact (UNEP, 1987). Waterlogging and salination lead to 200,000 to 500,000 hectares of irrigated land coming out of production annually (Goldsmith and Hildyard, 1984). The carrying capacity for human beings of all this land is essentially reduced to zero. When land deteriorates to this extent in poor countries, its inhabitants are forced either to join masses of displaced peasants in swelling urban slums or to migrate onto other marginal land where the cycle repeats itself.

Similarly, partly in response to population pressures, human beings are moving in ever-greater numbers into tropical moist forests (TMFs), where rapid deforestation and unsustainable agricultural practices render this land economically useless as well (Raven, 1988). Population growth among traditional shifting cultivators also threatens TMF by accelerating the slash-and-burn cycle to the point that the forest lacks time to recover between cuttings (Ehrlich et al., 1977).

Currently, only about half of Earth's original 16 million km^2 of TMF remains, and this is being severely disturbed (through intensive logging and slash-and-burn agriculture) or completely cleared at an annual rate of roughly 200,000 km^2 (Myers, 1988). Unless these patterns change, in 40 years relatively undisturbed TMF will be restricted to scattered fragments on steep hillsides and a few "islands" in Amazonia, the Congo basin, and Southeast Asia.

Such wholesale destruction of ecosystems reduces or eliminates services they once provided to people living both within and far from them. In addition to undergoing severe soil erosion, some badly deforested regions (where evapotranspiration is greatly reduced) suffer locally drier climates. In the Panama Canal area, as in some other tropical regions, there has been a steady decline in rainfall associated with the removal of most of the forest cover (Myers, 1988). Reduction in the recycling of water within the ecosystem may thus set up a positive feedback system that accelerates the loss of TMF.

The recent catastrophic flooding in Bangladesh can be attributed in part to massive deforestation in the Himalayas (Swaminathan, 1988), a phenomenon closely tied to population growth. The consequences of the loss of tropical biodiversity on carrying capacity are even more widespread. Industrialized countries rely heavily on tropical species for genetic material needed for the maintenance and improvement of strains of crops now in production (Myers, 1983) and for the development of new crops that could improve diets of human populations in the tropics (Ehrlich and Ehrlich, 1981).

Even more threatening than these direct effects of the human population on local and regional carrying capacities are human impacts that operate by changing global systems indirectly. The most important of these (but far from the only one) involves exacerbation of the natural long-term trend of interglacial warming.

GLOBAL WARMING

Anthropogenic climate change has been a matter of deep concern among environmental scientists for more than 2 decades (Bryson and Wendland, 1968; Ehrlich, 1968; Ehrlich and Ehrlich, 1970; SCEP, 1970; Ehrlich et al., 1977). The consensus among atmospheric scientists now is that the increased injection of greenhouse gases into the atmosphere due to human activities has already committed the planet to a warming of at least 1 or 2°C (Abrahamson, 1989). Furthermore, there seems to be little prospect of curbing future emissions sufficiently to prevent an average temperature rise of 3 to 4°C, or even more. To put this into perspective, consider that the average surface temperature of Earth during the last ice age was only 5°C cooler than it is today (Schneider, 1988)!

While the climatic effects of such a warming cannot be predicted with accuracy, computer models indicate that among the more likely results will be a decrease in water availability in the world's major grain belts. In addition, it is agreed that climate change will occur at a rate unprecedented in recorded history--possibly 10 to 50 times faster than the average natural rates of change following the last ice age (Schneider, 1988).

This degree and pace of change will inevitably cause major disruptions in world agriculture. Shifting climate belts will require major adjustments in irrigation and drainage systems at a cost of as much as $200 billion worldwide (Postel, 1987). Farmers will have to switch to drought-resistant crops where possible, thereby incurring reduced yields (drought-tolerant grains have an average yield less than half that of corn) (Brown, 1988). Drought-reduced harvests, like those of the late 1980s, can be expected to occur with greater frequency and severity.

Northward migration of temperature/rainfall belts that are favorable for grain production may at first glance appear beneficial to agriculture in regions like Canada and the northern part of the Soviet Union, where low temperatures and growing season frosts are limiting factors. But in many of those areas, thin, infertile soils will severely constrain productivity (Jenny, 1980).

Similarly, an increase in carbon dioxide concentration may enhance potential productivity, but it is doubtful that this will yield a net benefit in the face of so many other limitations. Higher temperatures and more carbon dioxide may unfavorably change relationships between crops and their pollinators, competitors, or pests. Finally, the unwillingness of governments to take many of the steps necessary to deal with nearly certain and unprecedented change will result in considerable delay and will exacerbate the socioeconomic problems involved in making adjustments.

Humanity has few options in making such adjustments to the projected greenhouse warming. The negative impact of climate change on global carrying capacity is not likely to be offset by increased agricultural yields through bringing more land into production or through increased fertilizer use. The potential for increasing the world's cultivated area is slim--the land area planted in grain worldwide has actually declined by about 7 percent since 1981 (Brown, 1988), due mainly to three changes: abandonment of deteriorated land; conversion of cropland to nonfarm uses, especially in densely populated regions; and set-asides in the United States.

The primary prospect for expanding food production thus rests with the potential for increasing yields through more intensive cropping, increased fertilizer use, or development of more productive strains. While it is still unclear how much higher yields can be raised (within economic constraints--inputs are limited by costs), no marked increases are foreseen in the near future as each of these avenues is approaching saturation under current economic conditions (Brown, 1988).

Global warming will also cause a rise in sea level due first to thermal expansion of the oceans and later to the melting of ice caps in polar regions, where the projected temperature rise is expected to be most dramatic. The predicted sea-level rise of as much as 1.4 to 2.2 meters by the end of the next century (Jacobson, 1988) will not only decrease food production through flooding of agricultural land, but will also displace millions of people from their homes and livelihoods. Damage to fisheries from inundation of wetlands that support them will adversely affect the nutrition of people who are heavily dependent on that food resource.

Coupled with land subsidence due to natural processes and the extraction of oil and groundwater, sea-level rises in some localities will be much higher than the average. Low-lying, fertile, and sometimes heavily populated deltas (e.g., the Brahmaputra/Ganges and Nile deltas) are likely to be submerged first. In Bangladesh and Egypt alone, an estimated 46 million people may be threatened by flooding (Jacobson, 1988).

Much larger areas of coastal land will become unsafe for human habitation because of the threat of storm surges carrying far inland. Developed countries, although more capable of resisting the rising seas, will not be immune. Holland may have to flood some of its reclaimed agricultural land with Rhine River water to prevent saltwater intrusion into groundwater supplies (Schneider, 1988). In Florida, much of the Everglades will be lost (with deleterious effects on fisheries), aquifers will be salinized, and large areas will be made much more vulnerable to storm damage. The increased frequency and severity of natural disasters (e.g., drought, storms, and flooding) associated with global warming, at a time when ecosystems are already stressed, will further reduce Earth's carrying capacity by decreasing the land area suitable for agriculture and human habitation.

An increased frequency of drought would also render food production less predictable, thereby reducing carrying capacity. Such reductions would be very serious, inasmuch as humanity is unable to feed itself adequately under current production and distribution systems.

A study by the Alan Shawn Feinstein World Hunger Program at Brown University (Kates et al., 1988) estimated that, even if food were equitably distributed (and nothing diverted to livestock), the all-time record food production of 1985 could have provided a minimal vegetarian diet to about 6 billion people, a number projected to be exceeded within the next decade. The same global harvest, allowing a diet with about 15 percent animal products, could feed some 4 billion people. A diet consisting of 35 percent animal products, similar to that consumed by most North Americans and West Europeans today, could be provided to only about 2.5 billion people--less than half of today's population. These estimates assume a 40 percent loss of the food harvested to pests and wastage before consumption, a Food and Agriculture Organization estimate that may be somewhat high. But even if that figure were 20 percent, it would not permit anything but an adequate vegetarian diet for today's population.

MODELING GLOBAL CHANGE AND FOOD SECURITY

To examine the possible effect of climate change on food production, we constructed a simple global model (for details, see G. C. Daily and P. R. Ehrlich, "An exploratory model of the impact of rapid climate change on the world food situation," in preparation) that simulates population growth, annual agricultural output, annual consumption, and the frequency and severity of unfavorable weather patterns such as occurred in 1988. The model determines the amount of food available for consumption (production plus carry-over stocks) in each year over a 20-year period. For all runs of the model, we assumed that average increases in grain production would keep up with population growth (1.7 percent annually). In years with favorable weather, we assumed that a surplus of 50 million metric tons of grain was produced. We then varied the frequency and severity of unfavorable weather patterns.

Under our most "optimistic" scenario, unfavorable climatic events occurred on average once every 5 years and caused a 5 percent reduction in grain harvest, roughly the magnitude of the climate-caused drop in 1988. Under our most "pessimistic" scenario, the mean time between unfavorable climatic events was 3.3 years, and each event caused a 10 percent drop in grain production below the trend.

In order to simulate the feedback between availability of food and population size, it was assumed that a food deficit of 1 metric ton of grain resulted in two incremental deaths. Roughly three people are supported by each ton of production now, but about one-third of all grain is fed to animals, so compensation is theoretically available by consuming more grain directly.

Actual death rates might, of course, be raised further than this indicates. In the real world, undernutrition occurs mainly among the poorest people, perhaps the bottom one-quarter or one-fifth of the population. This group bears the brunt of any deficits, while the rest usually can maintain adequate diets (although probably at higher prices). Because of the disproportionate burden on the poor, disease and hunger may take a heavier toll on them than our all-or-nothing simplification suggests.

Results of the model suggest that the optimistic scenario (a 5 percent reduction in grain harvest on average twice per decade) would not lead to complete depletion of world grain stocks, although world food security would be threatened. These reductions would have little effect on overall population growth. Under the pessimistic scenario (10 percent reductions on average three times per decade), however, severe deficits in grain stocks occur about twice per decade, each causing the deaths of between 50 and 400 million people.

Weather patterns that might cause such drops include, for instance, repeats of the 1988 North America/China drought event, with the same or greater severity, or totally different patterns involving other areas. In short, we have not incorporated the question of the pattern of crop failures that would lead to declines in grain production. We also have not considered compensatory actions such as bringing set-aside land in the United States back into production, conversion from feed to food crops, or the general intensification of agricultural activity that would result from increased demand for food, except to the degree they are subsumed in our "constant average increase" assumption.

We have also perhaps been pessimistic in not incorporating the possibility of increases in production because of technical innovations stimulated by famines. On the other hand, some of our implicit assumptions about carrying capacity are optimistic. We have not, for example, incorporated additional drops in harvest due to social breakdown related to famines, the spread of disease through malnourished (and thus immune-compromised) populations, or inappropriate aid programs that damage the agricultural sectors of recipient nations.

Indeed, most of our basic assumptions could be considered very optimistic. For instance, agricultural production is no longer keeping pace with population growth in Africa or Latin America. Furthermore, we have assumed that (climatic change aside) production can be kept growing for 2 decades more in spite of massive erosion of topsoil, increased waterlogging and salinization in irrigated areas, dropping water tables, deforestation leading to regional drought and flooding, desertification, accelerating conversion of land to nonagricultural uses, and so forth.

The model is, of course, simply an aid to thinking about the possible consequences if short-term climatic change were to cause drops in grain production of a magnitude roughly comparable to those known to have been caused before, and considering the rest of the system to be essentially "surprise free." Our results are not predictions; they are simply indications of the nature of problems that may occur if the global warming leads to an increased frequency and severity of climatic events deleterious to agriculture.

CONCLUSIONS

The population-food system has no "fail-safe" backup mechanisms designed into it, even if climates should remain very favorable to food production. We depend on the statistical "cushion" that adverse weather and unusual pest outbreaks do not occur everywhere at once. To the degree that global food production becomes more concentrated (as in North America), humanity becomes more vulnerable. There is no time to be lost in moving toward population shrinkage as rapidly as is humanly possible; the momentum of population growth ensures that human numbers cannot start to decline as a result of reduced fertility in less than half a century under any realistic assumptions.

Not only is population control required, but governments and societies must bend their efforts to reduce the rate of global climatic change (Ehrlich, 1988). The 1988 drought spurred the U.S. Congress into action, but bills first introduced in 1988 were still under debate in the spring of 1989 when the prospect of another drought-reduced harvest in North America seemed very real. Concerted action to start reducing the emission of greenhouse gases is needed now because (1) the resistance to implementing changes is so great and (2) the lead time on many effective actions will be a decade or more. The problem is especially acute, since leaders in the rich nations have largely failed to realize the magnitude of the changes necessary if the warming is to be significantly slowed. Action in rich nations is needed also to ensure that poor nations will have some chance to develop through use of their indigenous energy resources (Ehrlich and Ehrlich, 1989).

For the indefinite future, Homo sapiens will face major challenges in supplying everyone with adequate diets. Production must be increased while at the same time curbing the destruction of irreplaceable soils, overdrafts of "fossil" groundwater, and the destruction of biodiversity. Much more effort should go into reducing wastage of food between field

and stomach, strengthening the agricultural sectors of poor nations in
ways that promote their food security, and improving the equity of food
distribution. Even if all of these daunting ecological, economic,
social, and political tasks can be tackled simultaneously, there is no
guarantee of success.

Only one element of carrying capacity--food--has been examined in
this paper, and many of the complex interactions in the population-food-
climate complex have not even been explored. Nonetheless, our
preliminary analysis suggests that there is no room for complacency
whatsoever.

ACKNOWLEDGMENTS

This work is cosponsored by The Center for Conservation Biology of
the Department of Biological Sciences and the Morrison Institute for
Population and Resource Studies, Stanford University. We thank Susan
Harrison for helpful comments on the manuscript.

REFERENCES

Abrahamson, D. 1989. Global warming: the issue, impacts, responses.
In D. Abrahamson, The Challenge of Global Warming, in press.
Brown, L. 1988. The changing world food prospect: the nineties and
beyond. Worldwatch Paper 85. Worldwatch Institute, Washington, D.C.
Bryson, R.A., and W.M. Wendland. 1968. Climatic effects of atmospheric
pollution. Paper presented at AAAS Meeting, Dallas.
Ehrlich, P.R. 1968. The Population Bomb. Ballantine Books, New York.
Ehrlich, P.R. 1988. The global commons and national security. Paper
presented at the Symposium on Climate and Geo-sciences, May 1988.
Louvain-la-Neuve, Belgium.
Ehrlich, P.R., and A.H. Ehrlich. 1970. Population, Resources,
Environment: Issues in Human Ecology. W.H. Freeman, San Francisco.
Ehrlich, P.R., and A.H. Ehrlich. 1981. Extinction: The Causes and
Consequences of the Disappearance of Species. Random House, New
York.
Ehrlich, P.R., and A.H. Ehrlich. 1988. Earth. Franklin Watts, New
York.
Ehrlich, P.R., and A.H. Ehrlich. 1989. How the rich can save the poor
and themselves: lessons from the global warming. Paper No. 15.
Stanford Institute for Population and Resource Studies.
Ehrlich, P.R., A.H. Ehrlich, and J.P. Holdren. 1977. Ecoscience:
Population, Resources, Environment. W.H. Freeman, San Francisco.
FAO, UNFPA, and IIASA. 1982. Potential Population Supporting Capacities
of Lands in the Developing World. Technical Report of Project
INT/75/P13. Rome.
Goldsmith, E., and N. Hildyard. 1984. The Social and Environmental
Effects of Large Dams. Sierra Club Books, San Francisco.
Holdren, J.P., and P.R. Ehrlich. 1974. Human population and the global
environment. American Scientist 62:282-292.

Jacobson, J. 1988. Environmental refugees: a yardstick of habitability. Worldwatch Paper 86. Worldwatch Institute, Washington, D.C.

Jenny, H. 1980. Soil Genesis with Ecological Perspectives. Springer-Verlag, New York.

Kates, R.W., R.S. Chen, T.E. Downing, J.X. Kasperson, E. Messer, and S.R. Millman. 1988. The Hunger Report: 1988. Brown University, Providence, Rhode Island.

Myers, N. 1983. A Wealth of Wild Species: Storehouse for Human Welfare. Westview Press, Boulder, Colo.

Myers, N. 1988. Tropical deforestation and climatic change. Paper presented at the Symposium on Climate and Geo-sciences, May 1988. Louvain-la-Nueve, Belgium.

Postel, S. 1987. Stabilizing chemical cycles. In L. Brown, W. Chandler, J. Jacobson, C. Pollock, S. Postel, L. Starke, and E. Wolf, State of the World 1987. W.W. Norton, New York.

Raven, P.H. 1988. The cause and impact of deforestation. In H.J. de Blij, ed., Earth '88: Changing Geographic Perspectives, pp. 212-227.

Schneider, S. 1988. Doing something about the weather. World Monitor. December.

Study of Critical Environmental Problems (SCEP). 1970. Man's Impact on the Global Environment. MIT Press, Cambridge, Mass.

Swaminathan, M.S. 1988. Global agriculture at the crossroads. In H.J. de Blij, ed., Earth '88: Changing Geographic Perspectives, pp. 316-331.

United Nations Environment Programme (UNEP). 1987. Environmental Data Report. Basil Blackwell Ltd., New York.

Vitousek, P.M., P.R. Ehrlich, A.H. Ehrlich, and P.A. Matson. 1986. Human appropriation of the products of photosynthesis. Bioscience 36(6):368-373.

PART B   THE EARTH SYSTEM

4

THE EARTH SYSTEM

Digby J. McLaren

Our ideas about the earth have changed since James Hutton first gave us a model 200 years ago. He recognized the existence of an earth system and correctly outlined a model of ongoing change through small increments over an enormous time period. He thus paved the way for Darwin's still broader biological conceptions on the same basis. Hutton's model, however, was not evolutionary, and he really did make the oft-quoted remark, "We find no vestige of a beginning, no prospect of an end." It is ironical today, when we are at last approaching an approximation to a new model of the earth, that we are faced with the very real prospect of an end.

We now see the earth as a small planet in space that is inherently changeable. Its liquid core and mantle are heated by radioactive elements that still remain from its origins some 4.5 billion years ago. This heat induces ongoing crustal and mantle movement described under the general term "plate tectonics." Within this system there are many subsystems of change at present acting at different rates, some rhythmic and others episodic. As a consequence, it is beyond our capacity to predict future changes accurately. Some are manifest in earthquake and volcanic activity on land and at the ocean ridges. They are linked to change in the relative position of the plates leading, in turn, to changes in climate and ocean circulation and in the ambient life forms at or near the surface of the planet. All are ongoing and currently unpredictable.

The planet, with its life forms, is part of the solar system and is thus influenced by the sun and by variations in earth tilt and orbit round the sun. These induce further changes in atmospheric and ocean circulation, and therefore climate. Finally, the planet has been constantly bombarded by material in the form of meteorites and comets, some large enough to cause further massive changes in the earth system and its biomass. We are far from being able to tie all these variables together into a coherent model.

Life has played an important role in shaping the physical and chemical nature of the planetary surface. Life developed in balance with the changing environment as a result of an evolutionary process driven by those changes. In the very recent past, the emergence of the human race has begun to cause change in the environmental flux more rapidly than, and in a different manner from, the established system. With essentially free energy supplied by fossil fuels, our species has become, during the last 2 centuries, a dominant force for change on earth by any measure we may apply.

We are now able to chart past and current environmental changes, and techniques recently developed enable us to view the land, oceans, and atmosphere from space, and to measure secular changes in climate, cloud and ice cover, soil moisture, and marine and land bio-productivity. Ice cores have furnished an accurate record as far back as 160,000 years of global temperature, levels of atmospheric carbon dioxide, and variations year by year in wind-borne sediments, including volcanic events. Other techniques allow us to penetrate more deeply into the past.

Direct measurement may now be made of the accelerating effects of quarrying by humans of soils, forests, and ground water, encouraged by an economic system not yet adjusted to evaluating the sustainability of a resource. Quarrying here implies reducing the capacity of a resource, commonly assumed to be renewable, by overuse to such a degree that recovery will not occur on a human time scale, if at all.

The scientist may point out and measure many of the changes that are taking place in our immediate environment and in the time scale of our own lives. But it will take very much more than science to change our current system and persuade us to learn to live in balance with the earth's ecosystem. It seems to me that three major forces are at work to prevent this from happening. The following remarks are neither scientific nor social; they are the kind of observations that an intelligent martian, newly arrived on earth, might make, unhampered by axiom or belief.

1. Population growth seems to be the single most important factor in increasing environmental stress, including depletion of materials and energy resources and a runaway increase in solid, liquid, gaseous, and heat waste. We are currently adding 1 billion people to the population of the world every 11 years. And we must remember that most of the resources and most of the waste produced are due to the activities of a relatively small proportion of the total population. In certain areas of the world people are reproducing much more rapidly than in the West, but a Western baby will drive a car when he or she grows up and will use resources and produce waste at a rate 100 to 500 times greater than that of many of the babies being born in the Third World.

2. Nonrenewable energy consumption is another significant cause of environmental stress. Fossil fuel use continues to grow; as the use of petroleum products as fuel continues to decline relatively, it will be more than offset by an increase in coal burning and a concomitant increase in greenhouse gases and particulates in the atmosphere. The warming effect appears to have started: how much shall we accept, bearing in mind the inevitable change in climate and rise in sea level?

3. Global military expenditures now total nearly $1 trillion per year. In the light of problems facing the world today, these figures represent an enormous, almost unimaginable waste of resources and human ingenuity.

We are all in this together, and so we must find a joint solution. It is too late to build walls around or put roofs over regions of the world. The problems are exclusively global and the solutions must be also. We find we are now faced with a task that is more difficult than anything we have ever contemplated: to decide how we may continue to live on this small planet. For if we depart from ecological balance to the extent that we destroy most of the remaining life on earth--and the big killing is under way--then surely we are dooming ourselves to a similar fate. In other words, we must learn to live in balance with the world we find ourselves in.

The human being is an animal that has moved out of ecological balance with its environment. Humankind is a wasteful killer and despoiler of other life on the planet. This normal and apparently acceptable behavior is licensed by a belief in God-given resources and encouraged by an economic system that emphasizes short-term profit as a benefit and has not learned to put a real cost on the resources we consume.

I believe, therefore, that it is perfectly proper, as a scientist, to appeal for an inductive approach in looking at our present condition on earth, to draw empirical conclusions, including constructing some worst-case scenarios, and to attempt to assign probabilities to them. I began these remarks by focusing on the earth system, and I find that the human species has taken over. I suggest that we do not know enough to decide how to run this planet. We are forcing our will upon it, using the steadily depleting resources and increasing waste discharge, while at the same time claiming that we must aim for sustainability. May we attain this globally, or have we passed the limits within which it may be achieved?

5

MISSION TO PLANET EARTH REVISITED*

Thomas F. Malone and Robert Corell

Environmental scientists are blazing new, bold, and imaginative
trails to discover the interactions that bind the elements of land,
water, air, biota, and energy into planet Earth. Understanding these
interactions is imperative if future generations are to inherit a
habitable planet, because human activity has expanded and developed to
the point where anthropogenic environmental changes are jeopardizing
continued human existence. Science stands at the threshold of an
unprecedented opportunity to study and learn the far-reaching implications of both anthropogenic and natural environmental changes.

A comprehensive study of the global environment is within reach.
Such a study would give humanity the knowledge base necessary to
intervene on its own behalf--to reverse the trends of global environmental degradation and to bequeath to future generations a benign Earth.
The complex and vital nature of this endeavor demands careful delibera-
tion. A rationale for such a study, titled "Mission to Planet Earth,"
was set forth in _Environment_ two years ago.[1] It rested on five central
considerations:

   o A revolution under way in the sciences is leading to the treatment
of the physical, biological, chemical, and geographical parts of the
global system as an integrated and responsive whole.
   o Scientific understanding of each part and the interactions among
the parts is approaching a stage at which describing the controlling
physical, chemical, and biological processes in quantitative form (i.e.,
with mathematical models) will be possible.
   o Scientists are developing with unprecedented speed the necessary
technological tools (in situ and space-based measurements, computers,
and communications) for a holistic analysis of Earth's system. These
tools could enhance the ability of science to predict changes in the
environment.

---

*This paper, which was available at the forum, is reprinted, with
permission of the Helen Dwight Reid Educational Foundation, from
Environment 31(3):6-11, 31-35 (April 1989). Published by Heldref
Publications, 4000 Albemarle St., N.W., Washington, D.C. 20016.
Copyright (c) 1989.

o With exponential growth in population, agriculture, and industry, human activity is becoming a powerful factor, or forcing function, for global change.

o The capacity of the global life-support system to sustain a technologically advanced and exponentially expanding civilization is likely to collapse within the foreseeable future.

The rationale for a "Mission to Planet Earth" has been significantly transformed during the past two years. Opportunities to respond to this scientific and technological revolution are still open and have, in fact, expanded on their own merits. The compelling need now, however, is to increase the knowledge base underlying major policy decisions on societal and national behaviors that affect global well-being. A new era, characterized by a grand convergence of natural sciences, social sciences, engineering, public policy, and international relations, is emerging.

Many national and international research organizations have incorporated the "Mission to Planet Earth" rationale into their language and programs. At a meeting in Berne, Switzerland, in September 1986, the General Assembly of the International Council of Scientific Unions (ICSU) decided to establish the International Geosphere-Biosphere Programme: A Study of Global Change (IGBP). Since that meeting, organization and planning have proceeded with remarkable speed. IGBP represents a herculean effort by scientists worldwide to give humanity the knowledge base to fashion policies that will reverse the global environmental decline.

GLOBAL CHANGE IN THE LIMELIGHT

What has pulled scientists out of their separate laboratories to call for and organize this effort? Several recent developments have pushed global change and environmental policy issues on center stage. One such development is the frequent scientific prediction of a rise in global temperature of several degrees centigrade as a result of increased emissions to the atmosphere of greenhouse gases, which trap long-wave radiation emanating from Earth's surface. The possibility that human activity, especially fossil fuel use, could exacerbate the atmosphere's greenhouse effect was recognized as long as a century ago. Not until a 1985 meeting of Earth scientists in Villach, Austria, however, did the scientific community become sufficiently impressed with all the evidence for the probable magnitude of the greenhouse effect to call for political action. (For a discussion of the conclusions drawn at the conference in Villach, see Jill Jäger's "Climate Change: Floating New Evidence in the CO_2 Debate" in the September 1986 issue of Environment.) In the summer of 1988, James Hansen, a scientist with the National Aeronautics and Space Administration (NASA), intensified the policy debate while testifying before the U.S. Congress by implying that the first sign of greenhouse warming had already been detected in the climate records.[2]

Another development that has pushed global change into the limelight is the sharp seasonal decrease of the stratospheric ozone layer over the

Antarctic and signs of incipient decline over the Arctic. The ozone
layer screens out the sun's harmful ultraviolet radiation. The warning
by scientists in 1974 that chlorofluorocarbons (CFCs) drifting up into
the stratosphere constitute a threat to the ozone layer turned out to be
remarkably prescient. By 1987, the evidence for ozone depletion was so
persuasive that international agreement was quickly reached on the
Montreal Protocol to limit the production of CFCs. Already this year, a
number of countries have announced plans to accelerate the protocol's
implementation by banning certain CFCs entirely by the year 2000.

A third development is the annual loss of more than 10 million
hectares of forest cover in the tropics. In industrialized countries,
millions of hectares of forest are destroyed each year by fire, and many
more millions of hectares are jeopardized as a result of acid deposition.
Worldwide, the rate of forest loss is about 1 acre (or 0.4 hectares) per
second. At the same time, more than 5 million hectares of new desert are
formed each year as a consequence of land mismanagement in semi-arid
regions. Finally, the rate of extinction of plant and animal species has
reached alarming and probably irreversible levels. This loss is robbing
future generations of valuable resources for food, industry, and medicine. All of these developments have helped to heighten the public's
awareness of global change.

GROWING INTEREST

Scientists have repeatedly verified the manifestations of global
change.[3] Plans to cope with these disastrous changes are being formulated worldwide.[4] Thus, research like that proposed in "Mission to
Planet Earth" has been elevated to new importance by a greater perception on the part of the public and policymakers that human activity is
rapidly approaching a level at which human-induced change of the global
environment will be on a scale equivalent to change produced by natural
forces. Some indications that there will be definite winners and losers
in the global change game have sparked additional enthusiasm for worldwide conservation and research. More fuel for this fire has been the
perception that the human carrying capacity of Earth may soon be
stressed to a point where catastrophic consequences should be expected in
regions currently characterized by high population density and growth.

The very threat of such consequences has raised issues of social
equity and international security, and a stirring of political will for
environmental causes is evident:

o In the United States, two dozen bills concerning the environment
were cosponsored by more than 400 senators and representatives during the
100th Congress (1987 and 1988).

o In his 1988 campaign for the U.S. presidency, George Bush promised
to convene an international conference on global warming during his first
year in office.

o Last fall, in a speech to The Royal Society, Great Britain's Prime
Minister Margaret Thatcher labeled protection of the balance of nature as
one of the "great challenges of the late twentieth century."[5]

o At their December 1987 summit meeting in Washington, D.C., General Secretary Mikhail Gorbachev and President Ronald Reagan agreed to a collaboration on issues of climate and environmental change.

o In his December 1988 address to the United Nations General Assembly, Gorbachev remarked that "international economic security is inconceivable unless related not only to disarmament but also to the elimination of the threat to the world's environment."[6]

o In his 1988 World Environment Day message, Mostafa Tolba, Executive Director of the United Nations Environment Programme (UNEP), warned that "it may take another 15 years before scientists can give reliable predictions of what warming will mean in each region. But by then it may be too late to act."[7] He called on political and industrial leaders to cooperate with one another and with climate scientists to finance more international research and coordination that will produce more information more quickly.

Last year, greater attention to climate change issues by the news media was both a cause and effect of the new public awareness. The National Geographic Society devoted the December 1988 issue of its magazine to the theme "Can Man Save This Fragile Earth?"[8] Time magazine broke tradition and set aside identification of the man or woman of the year to dedicate its New Year's issue to "Planet of the Year: Endangered Earth."[9] Beyond just diagnosing the ills of the planet, Time proposed a 19-point action agenda for all nations and 8 steps that the United States should take to address Earth's environmental crisis. In fact, three major U.S. news magazines--Time, Newsweek, and U.S. News and World Report--devoted cover articles to global change during 1988.[10] More recently, the very critical nature of global change was well summarized by the U.S. National Academy of Sciences' recommendations to the Bush administration.[11] (Excerpts of these recommendations may be found on page 30 of the January/February issue of Environment.)

Growing public interest in global change issues has already prompted some political action. For example, in April 1987, the World Commission on Environment and Development analyzed and explicated these issues in Our Common Future, the report commissioned by the United Nations General Assembly.[12] The commission argued persuasively that the issues of human and economic development, environmental quality, and natural resource husbandry are highly interdependent; sound development requires a sound environment and a strong natural resource base, and an unhealthy environment and resource depletion ensure poor development. The commission urged expansion of the role of the scientific community in planning, decisionmaking, and implementing measures for coping with climate change. (For a review of Our Common Future, see the June 1987 issue of Environment.)

In September 1987, several member countries of UNEP met in Montreal and signed an agreement to stem CFC production. Although the Montreal Protocol is very modest (and many say insufficient), the fact that a precedent was established in arriving at a multilateral agreement somewhat offsets the disappointment of the compromises that became necessary to enhance the protocol's appeal to prospective signatories. Encouraged by this step, UNEP and the World Meteorological Organization established the Intergovernmental Panel on Climate Change to address the vastly more difficult problem of greenhouse warming by assessing the science, impacts, and policy implications of that topic. (For more on this panel, see the January/February 1989 issue of Environment.)

In the summer of 1988, 300 experts in science, law, environment, and economics met at the World Conference on the Changing Atmosphere: Implications for Global Security in Toronto, Canada. The Conference Statement concluded that the gravity of the risk of global warming called for a 20 percent reduction in carbon dioxide emissions from 1988 levels by 2005.[13] It urged support for such efforts as the World Climate Programme, the International Geosphere-Biosphere Programme, and Human Response to Global Change. (For more on the conference, see a special report on the Conference Statement in the January/February 1989 issue of Environment and the statement's 22-point "Call for Action" on page 31 of the September 1988 issue of Environment.)

A SCIENTIFIC INITIATIVE

In September 1986, an ad hoc planning committee recommended to the General Assembly of ICSU that an International Geosphere-Biosphere Programme be organized to guide and assess scientific research on global change. ICSU agreed to the program and, in early 1987, appointed a Special Committee to plan an IGBP that would "study the progressive changes in the environment of the human species on this Earth, past and future; to identify their causes, natural or man-made; and to make informed predictions of the long-term future and thus of the dangers to our well being and even to our survival; and to investigate ways of minimizing those dangers that may be open to human intervention."[14] To realize these goals, IGBP will sponsor research in several critical areas and will actively support other research programs, both national and international.

During the first half of 1988, the Special Committee developed a preliminary research plan.[15] In October, the IGBP Scientific Advisory Council, composed of members of ICSU, national IGBP committees, and other organizations from 40 nations, met in Stockholm to review the research plans. The Special Committee proposed a broad array of activities to the nearly 200 scientists assembled. In the end, five broad topics under which research could be grouped and coordinated were defined: terrestrial biosphere-atmosphere chemistry interactions; marine biosphere-atmosphere interactions; biospheric aspects of the hydrological cycle; effects of climate change on terrestrial ecosystems; and global analysis, modeling, and interpretation.[16]

Certain research areas deserving concentrated attention were identified, including data and information systems, geosphere-biosphere observatories, and techniques for extracting environmental data of the past.[17] IGBP task forces are focusing on each of these areas to develop a research agenda for the 1990s. For each focus, they are investigating an array of interrelated research projects either proposed or already under way (e.g., World Climate Research Program, World Ocean Circulation

Experiment, Man and the Biosphere Program, International Global Atmospheric Programme, and the International Satellite Cloud Climatology Project).

Some of the most exciting research topics opening up include Earth history, the forcing functions of global change, Earth-system fluxes, and predicting resource availability. By documenting historical Earth processes and studying their results and by assessing present and anticipated anthropogenic impacts, future global changes might be predicted more accurately. Theories and models of future global change can be tested by comparing them to actual historical changes. Thus, Earth history will be exploited to identify the forces behind climate change and to investigate the coupling between biogeochemical processes and the physical climate (e.g., the connection between greenhouse-gas emissions and global warming). This work will require an extensive observation system to document past Earth processes and future environmental change.

Forcing functions of global change, including solar and orbital changes, solid-Earth processes, and, in particular, human activities that influence the Earth system on a planetary scale, must be analyzed and understood. Variations in solar activity influence climate change over scales of decades, centuries, and millennia. The solar energy flux can affect not only the physical environment but also such biological processes as photosynthesis and respiration. Solid-Earth processes such as volcanic eruptions and marine vents affect the global climate and may cause extinctions.

Changes in a human activity like land use can cause global change by disturbing carbon storage, nutrient cycles, the hydrologic cycle, atmospheric composition, and the reflection of solar energy from the Earth's surface. As global population grows and humans convert more and more natural resources into goods and services, anthropogenic perturbations of the environment can only increase. (Figure 5.1 shows projected population growth through the year 2120.) The rate of change itself may be a forcing function because the rate of change can affect the kind of change; short-term but extreme perturbations of Earth's system can cause more dramatic changes than do long-term, gradual perturbations.

Scientists must pay more attention to the interactions of physical, chemical, and biological components of the system and to the flux of energy, water, and chemicals throughout all of Earth's domains, instead of concentrating on pieces of the system as if they were static and isolated. Understanding these interactions is critical in part because global change is nonlinear: It occurs as a threshold response to a continual force, just as the back of the proverbial camel breaks suddenly with the addition of one more straw. Moreover, the change effected upon the atmosphere by the ocean will in turn affect the ocean (as well as other domains). Special attention must be paid to the cycling of chemicals (carbon, nitrogen, sulfur, phosphorus, and trace gases) through the physical, biological, industrial, and agricultural systems. A network of geosphere-biosphere observatories in selected ecosystems is envisioned to serve as regional research and training centers. As understanding of Earth processes is enhanced, attention will be turned

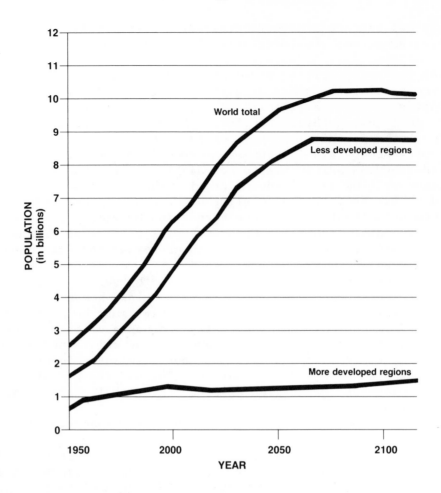

FIGURE 5.1 Projected population growth for developed and less developed regions through the year 2120.

Notes: More developed regions include Europe, North America, Australia, Japan, New Zealand, and the USSR. Less developed regions include Africa, Asia, Latin America, and Oceania.
SOURCES: Department of International Economic and Social Affairs, World Population Prospects as Assessed in 1980, Population Studies No. 78 (New York: United Nations, 1981); and Department of International Economic and Social Affairs, Long-Range Population Projections of the World and Major Regions, 2025-2150, Five Variants, as Assessed in 1980, 1981 (New York: United Nations, 1981).

to developing quantitative models capable of projecting global change into the future.

Global change will cause important, large-scale modifications in the availability and distribution of renewable and nonrenewable resources. Although the force behind the change may be global in scale (like global warming), predictions of resource availability are needed on a regional scale. (In the January/February issue of Environment, Thomas E. Graedel presents a methodology for assessing and predicting regional impacts of global change.)

OTHER INITIATIVES

An important contribution to global-change research planning was the massive report Earth System Science released in January 1988 by the Earth System Sciences Committee of the NASA Advisory Council.[18] The committee's mandate was to define a comprehensive, integrated program to obtain a scientific understanding of the entire Earth system and of the functions and interactions of its component parts. Such understanding could enable scientists to predict both natural and anthropogenic global changes over time scales of decades to centuries. The committee made detailed recommendations on five substantive topics:

o space-based and in situ long-term measurement of the global variables that define the vital signs of the Earth system and control its changes;
o fundamental description of Earth and its history;
o process studies and research focused on key Earth-system problems;
o development of Earth-system models to integrate data sets, guide research programs, and simulate future trends; and
o development of an information system to facilitate data reduction, data analysis, and quantitative modeling.[19]

The committee identified two distinct phases of work: near-term (1987-1995), to include the currently planned space missions and the conduct of essential process studies, and long-term (1995 and beyond), to deploy a new generation of space technology integrated with ground-based measurements to constitute a comprehensive Earth Observing System (EOS).

During April 1988, senior officials from 17 national space agencies gathered in Durham, New Hampshire, for the ISY Mission to Planet Earth Conference,[20] which was convened in connection with the International Space Year (ISY) planned for 1992. The conference established a Space Agency Forum for ISY and chose to make "mission to planet Earth" a major theme of ISY. Recently, the increasing number, diversity, and sophistication of space-agency Earth observation missions have underscored the importance of standardizing their output and making it readily accessible. Therefore, particular support was given to a proposal to mount a Global Information Systems Test (GIST) to develop globally accessible formats for data collected by national systems on two key problems: early detection of the greenhouse effect and deforestation.

The IGBP initiative at the international level is being supported by imaginative proposals emanating from many of the 70 scientific organizations adhering to ICSU.[21] For example, last year the Committee on Global Change of the U.S. National Research Council[22] recommended that U.S. contributions to IGBP include the development of an integrated EOS with space- and ground-based observatories and proposed initial research priorities that included studies of:

o water, energy, and vegetation interactions, to develop models of the coupling between climate and terrestrial ecosystems;

o fluxes of trace gases and nutrients between terrestrial ecosystems, the atmosphere, and the oceans;

o biogeochemical dynamics in the ocean, to understand and predict the effects of global change on ocean biogeochemical cycles and their feedback effects on global change;

o Earth history, to construct models of past climate change that could stand as a basis for validating models of future global change; and

o human interactions with global change, with special attention to land-use changes that affect both physical and biological parameters and to the residues from industrial/agricultural processes that perturb the global environment.[23]

EARTH OBSERVING SYSTEM

In the past, revolutionary advances in science have followed the development of an instrument that enables scientists to observe in detail some aspect of the natural world. For example, the invention of the microscope and its sophisticated progeny opened the fields of cellular and molecular biology and gave humans new understanding of life processes. No single piece of hardware triggered the current revolutionary shift in Earth sciences.[24] However, remote sensing from spacecraft together with the communications and computation capability now available have stimulated a holistic view of this planet and its atmosphere, oceans, land, fauna, and flora.

Much of the research described above requires the development and implementation of an extensive and elaborate Earth Observing System. EOS should identify and document past, present, and future global changes with both ground- and space-based sensors. The potential of space-based sensors is not yet appreciated widely enough, but the emerging capability has profound implications for the revolution under way in Earth sciences.

The data provided by EOS must be made accessible to researchers through a global information system incorporating the latest advances in communication and computers. In the same vein, internationalization of space-based observations of Earth's vital signs is urgent. The very modest first steps taken last April by 17 national space agencies to make "mission to planet Earth" a major theme of ISY 1992 should be expanded and institutionalized as a free-standing program, independent of the competing demands for resources, to advance the broad objectives of space science and technology.

THE U.S. GOVERNMENT RESPONDS

In close coordination with the National Research Council and ICSU, a U.S. strategy for global change research was developed by federal agencies and transmitted to Congress by the Director of the Office of Science and Technology Policy in the Executive Office of the President.[25] The strategy document accompanied President Reagan's budget message for fiscal year 1990, which begins October 1, 1989. The budget message proposed funding global change research with $190.5 million--an increase of 41 percent.[26] The U.S. Global Change Research Program aims to provide a sound scientific basis for national and international policy decisions on global change issues. The program's scientific objectives are to monitor, understand, and, ultimately, predict global change. The strategy document identified seven integrated and interdisciplinary elements of the program:[27]

o **Biogeochemical dynamics.** The study of the sources, sinks, fluxes, and interactions among the mobile biogeochemical constituents within the Earth system and their influences (including global warming) on the life-sustaining envelope of the Earth.

o **Ecological systems and dynamics.** The study of how aquatic and terrestrial ecosystems both affect and respond to global change.

o **Climate and hydrologic systems.** The study of the physical processes that govern the climate and hydrological systems central to global change, including the atmosphere, hydrosphere (oceans, surface and ground water, etc.), cryosphere (frozen regions), land surface, and biosphere.

o **Human interactions.** The study of the interface between natural processes and human activities. (The global environment is a crucial determinant of the human capacity for sustained development.)

o **Earth-system history.** The study of past natural environment change as it is revealed in rocks, terrestrial and marine sediments, glaciers and ground ice, tree rings, geomorphic features (including the record of changes in sea level), or other manifestations of past environmental conditions. As past analogues of possible future global changes, the records contribute to the understanding of the present Earth system, to the discrimination between natural and anthropogenic change, and hence to the prediction of future global change.

o **Solid-Earth processes.** The study of solid-Earth processes that affect the life-supporting characteristics of the global environment and especially those processes that take place at the interfaces between the solid Earth and the atmosphere, hydrosphere, cryosphere, and biosphere.

o **Solar influences.** The study of variability in solar brightness and its impact on atmospheric density, chemistry, dynamics, ionizations, and climate.

This strategy document sent to the U.S. Congress represents a crosscutting review and integration by the Office of Management and Budget of a number of initiatives by individual agencies with different purposes and characteristics in support of a national objective. This effort followed a procedure proposed by a committee of the presidents of the National Academy of Sciences, National Academy of Engineering, and Institute of Medicine and several of their respective councilors.[28]

KEY ISSUES

The attractive opportunities during the next decade for a true partnership between the scientific community and government to study global change are matched only by the challenges that must be met and overcome. For example, a balance must be achieved between the traditionally cautious scientist who tries to satisfy a seemingly insatiable appetite for information before supporting action and the sometimes overzealous environmental activists who maintain that deferred action only leads to more difficult decisions in the future. This dilemma can be resolved only by the wisdom and good judgment of political leaders. For instance, many actions that have not been proven necessary on purely scientific grounds are nonetheless advisable on economic or other grounds; increasing energy efficiency is a perfect example. Prudence alone suggests that such actions are desirable.

Redesigning Institutions

Another challenge is to reconstruct the present international institutional framework for addressing both research and policy; this framework was formed before the interdependence of nations became so apparent and when it was believed that global issues (food, weather, energy, and socioeconomic development, etc.) could be compartmentalized and addressed in relative isolation. National institutions for space research must be "internationalized"--that is, space agencies must develop closer interaction despite past competitiveness and political differences. The institutional framework must become capable of addressing issues that cut across disciplines and intermingle science and policy; at the same time, the framework must gain the support of nations with conflicting ideologies and in various stages of socioeconomic development. In light of characteristic institutional inertia, the fight for reform and renewal will be difficult.

Related to the issue of international institutional arrangements, but an urgent matter in its own right, is the required coordination among the wide array of research efforts directed toward each particular aspect of global change. Harmonious orchestration of these efforts is imperative; however, intrusive management of research must be avoided and the essential function of the individual investigator must be protected.

Social Sciences and Engineering

Another key issue is the involvement in global change research of disciplines not strictly in the domain of natural science but still relevant to understanding interactions between humans and their environment. For example, because some of the roots of global change are found in the metabolism of the industrial/agricultural system, it is important for the engineers and managers of those systems to participate actively in charting the future course.[29] Because global change arises from social systems as much as from physical, chemical, and biological systems, the full participation of social scientists is also urgent.

Although in the past many natural scientists resisted the inclusion of other disciplines in their research programs, during these last two years, the ubiquitous role of social and behavioral sciences and engineering in the study of global change has been widely recognized.[30] In its full flowering, IGBP will become a sustained, international, and coordinated research program to illuminate the interaction of the physical, chemical, biological, and social systems that regulate Earth's unique environment for life.

As IGBP evolves and matures, it should become an admirable collaboration among the scientific community, nongovernmental and intergovernmental organizations, and sovereign nations. The program increasingly will attract the interest of engineers and social and behavioral scientists. Their involvement should suggest new dimensions for research, such as assessing the societal impact of global change in all its myriad manifestations; analyzing possible public policies that should be considered to obviate certain kinds of global change, mitigate others, and adapt to still others; and developing policy options flexible enough to incorporate uncertainty, with respect to both the human contributions to global change and the unequal division of the positive and negative consequences of global change over local, regional, national, and international territories. An entirely new mode of interdisciplinary cooperation among natural scientists, social scientists, and engineers will be required. A new social contract must be drawn up among science, engineering, and society.

Developing Countries

One major challenge is to ensure the full involvement of developing nations in IGBP. A particular opportunity for them is afforded by the proposal to establish strategically located geosphere-biosphere observatories dedicated to training and research. It would be tragic if the use of high technology in space-sensors, communications, and computers precluded the participation of scientists from developing countries. The global policy issues that must be resolved in the years to come will require the support of every nation. The most effective way to ensure this support is to make specific provision for their participation in the development of the knowledge base that will undergird those decisions. In this way, each nation will realize the necessity of and work toward a convergence of international interests and aspirations. The magnitude of the task of unifying policies in developing and developed countries was revealed at the recent meetings in New Delhi and London. These meetings underscored the need to cooperate to establish the baseline that will make the difficult decisions tractable. Indeed, more and more, scientists are realizing that the single, indivisible Earth system belongs to one indivisible world in which it is insufficient to study one society in isolation from all others.

Financing

Financing IGBP and other such international research programs is another great challenge. If major decisions on public policy in response to prospective global change are to be solidly based on the best available information rather than on popular and political perception, creative solutions to finance the needed research will have to be fashioned.

Material needs fall into two categories: national financing of national programs and national financing of integrated international activities. Each category can be further divided into financing of the immediate planning phase and financing of the longer research phase.

Currently, it appears that the international preparatory effort requires anywhere from U.S.$1 to 2 million annually. If the U.S. funding for the planning phase is used as a guide, the ratio of domestic to international funding is greater than 100 to 1. The cost of the material resources for the research phase will probably be an order of magnitude larger for both the national and international programs. Still, even the cost of the research phase will seem small when compared to the costs of whichever measures are finally chosen to adapt to or influence global change. Clearly, new ground will have to be broken in integrating and coordinating national and international planning, research, and operations.

Financing work in developing countries is a special issue. Nations unable to finance national research could easily be left behind even though their participation in policy decisions is vital. New financing systems patterned after the International Foundation for Science in Stockholm and the International Development Research Centre in Ottawa, for example, will have to be arranged. Now is a good time to explore the feasibility of an International Science Foundation to fulfill at the international level the same need that was perceived at the national level by President Roosevelt and Vannevar Bush in the late 1940s and gave rise to the U.S. National Science Foundation. An institutional framework of this kind would represent an expansion of the activities supported by the International Foundation for Science and the International Development Research Centre.

CONTINUING QUESTIONS

The new public interest in global change raises some important questions. Can a global strategy to survive the upcoming changes be fashioned in a world of more than 150 sovereign nations at various levels of dynamism and socioeconomic development? Could the United Nations be given the power to police the global atmosphere as was proposed at the Netherlands summit meeting of 24 nations in early March? Will industrialized nations agree to "compensatory financing to Third World countries unable to bear the cost of the conservation and antipollution measures" needed?[31] Is the knowledge base adequate for an international program of action? Is "mission to planet Earth" simply a catch phrase, or does it constitute a viable focal point for transforming the scientific and technological triumphs of this century into a service, rather than a threat, to society? How can priorities be established when demands for resources are unlimited?

There are no easy answers to these and other questions that have surfaced in the past two years. Although many questions inevitably will remain unanswered for some time, certain vital steps must be taken at once. Environmental changes with profound consequences are impelling nations to seek politically and economically acceptable solutions. To be viable and effective, the solutions need to have broad authorship, because nations that participate in the derivation of a solution will be more likely to support and implement it. The Intergovernmental Panel on Climate Change is a promising forum for the needed effort because it has been structured to develop objective assessments of environmental impacts and to design response strategies based on an international consensus of scientific knowledge. The nations of the world should give this panel full support and hold its leadership to the highest standards of performance. One vital strategy, "parallel action," will ensure that scientific assessments and response strategies can proceed simultaneously.

Through IGBP, an agenda for informed debate, discussion, and action is emerging that warrants sustained attention during the approximately 4,300 days that remain before we cross the threshold into the third millennium. The implications of the IGBP research effort for science and society are profound. The next major milestone will be a June 1989 meeting in Brussels that will bring together the Special Committee and national IGBP committees to develop a synthesis of the various plans that were presented by the Special Committee and several national committees in Stockholm last October.

The global changes clearly visible on the horizon are rooted in the scientific and technological advances that have unlocked many of the secrets of matter, energy, life processes, and information and made this knowledge accessible for human purposes. The options for society are three:

o permitting civilization to be snuffed out by savaging the global environment with the weaponry scientific knowledge has made available;

o allowing the global environment and civilization to be gradually suffocated by exponential population growth and by uncontrolled and inequitable transformation of natural resources into the goods and services that sustain and give meaning to life; and

o planning and constructing a more prosperous, just, and secure world.

The coming decade will be one of the more critical periods in the several million years of human evolution. If humans are to survive safely the changes that clearly lie ahead, each day must be marked by discrete progress toward a better world. Knowledge, the coin of scientific enterprise, is the sine qua non of such progress. However, the first step perhaps is to "reaffirm a robust faith in the destiny of man."[32] It is the unique privilege and challenge of this generation to open this window of opportunity into that better world.

48

NOTES

1. Thomas F. Malone, "Mission to Planet Earth," Environment, October 1986, 6. Recently this title received broader public attention through the study by astronaut Sally Ride, Leadership: America's Future in Space (Washington, D.C.: National Aeronautics and Space Administration, 1987). The phrase "mission to planet Earth" has become a catch phrase describing the efforts to understand and respond to global and climatic change.
2. U.S. Congress, Senate Committee on Energy and Natural Resources, Greenhouse Effect and Global Climate Change, 100th Cong., 1st sess., pt. 2, 39.
3. Particularly relevant articles may be found in Mosaic 19 nos. 3/4 (1988); Earth System Sciences Committee, NASA Advisory Council, Earth System Science; A Closer View (Washington, D.C.: National Aeronautics and Space Administration, 1988); and the U.S. National Research Council, Space Science in the Twenty-First Century: Imperatives for the Decades 1995 to 2015: Mission to Planet Earth (Washington, D.C.: National Academy Press, 1988).
4. For U.S. planning see U.S. National Research Council, Committee on Global Change, Toward an Understanding of Global Change: Initial Priorities for U.S. Contributions to the International Geosphere-Biosphere Program (Washington, D.C.: National Academy Press, 1988). An account of international planning is found in J. J. McCarthy, chairman, The International Geosphere-Biosphere Programme: A Study of Global Change (IGBP) A Plan for Action, Report No. 4 (Stockholm: IGBP Secretariat, Royal Swedish Academy of Sciences, August 1988).
5. Christine McGourty, "Margaret Thatcher's U-turn on Support of Basic Research," Nature 338(6 October 1988):484.
6. Translated by the Soviet Mission to the United Nations, New York Times, 8 December 1988, A16.
7. Mostafa Tolba, "Global Warming: Window of Opportunity," speech delivered in Bangkok, 5 June 1988.
8. "Can Man Save This Fragile Earth?" National Geographic, December 1988.
9. "Planet of the Year: Endangered Earth," Time, 2 January 1989.
10. Ibid.; "The Greenhouse Effect," Newsweek, 11 July 1988; and William F. Allman, "Rediscovering Planet Earth," U.S. News and World Report, 31 October 1988, 56.
11. National Academy of Sciences, National Academy of Engineering, and the Institute of Medicine, The Four White Papers for the Transition Team (Washington, D.C.: National Academy Press, 1989).
12. World Commission on Environment and Development, Our Common Future (Oxford and New York: Oxford University Press, 1987).
13. H. L. Ferguson, conference director, Conference Statement, World Conference on the Changing Atmosphere: Implications for Global Security, Toronto, Ontario, Canada, 27-30 June 1988, published by Environment Canada.
14. Sir John Kendrew, president of the International Geosphere-Biosphere Programme, cited in McCarthy, note 4 above, page 3.
15. McCarthy, note 4 above.

49

16. Ibid.
17. Ibid.
18. Earth System Sciences Committee, NASA Advisory Council, Earth System Science: A Program for Global Change (Washington, D.C.: National Aeronautics and Space Administration, 1988).
19. Ibid.
20. H. Myerson, ed., Report of the ISY Mission to Planet Earth Conference: A Planning Meeting for the International Space Year (Washington, D.C.: US-ISY Association, 1988).
21. V. M. Kotlyakov, J. R. Mather, G. V. Sdasyuk, and G. F. White, "Global Change: Geographical Approaches (A Review)," Proceedings, National Academy of Sciences 85 (August 1988):5986-91.
22. U.S. National Research Council, note 4 above.
23. Ibid.
24. Earth System Sciences Committee, NASA Advisory Council, note 18 above.
25. Committee on Earth Sciences, Federal Coordinating Council for Science, Engineering, and Technology, Our Changing Planet: A U.S. Strategy for Global Change Research (Washington, D.C.: U.S. Government Printing Office, 1989).
26. Ibid.
27. Ibid.
28. National Academy of Sciences, National Academy of Engineering, and the Institute of Medicine, Federal Science and Technology Budget Priorities: New Perspectives and Procedures (Washington, D.C.: National Academy Press, 1988).
29. John Helm, ed., Energy: Production, Consumption, and Consequences (Washington, D.C.: National Academy Press, forthcoming).
30. H. K. Jacobson and C. Shanks, Report of the Workshop on an International Social Science Research Program on Global Change at the Institute for Social Research, University of Michigan, Ann Arbor, Michigan, 1987; and U.S. National Research Council, Committee on Global Change, note 4 above.
31. Edward Cody, "Global Environmental Power Sought," Washington Post, 12 March 1989, A27.
32. Pierre Teilhard de Chardin, Building the Earth (Wilkes-Barre, Penn.: Dimensions Books, 1965).

6

HISTORICAL PERSPECTIVES: CLIMATIC CHANGES
THROUGHOUT THE MILLENNIA

John E. Kutzbach

We are here to consider the prospects of global change and our common future. Our aim is to look forward. I want to share with you some perspectives about global change that we can gain from first looking backward to our common past. Why is it important to look backward first? The first reason is that the global changes that may occur in the next century may be larger than any changes that have occurred in recent centuries. We need to look to more distant times for examples of large global change. We will see that large global climatic changes have had large effects on plants, animals, and humans.

The second reason for looking to the past is that the past is a laboratory in which we can study these global processes and develop our predictive capabilities. If we can identify the factors that have caused global change in the past, and if we can successfully estimate past climates and climatic changes using our computer models, then we will gain confidence in our ability to estimate and anticipate future changes.

I will describe five examples of large global change from the past: one from about a billion years ago, another from several hundred million years ago, and others from one million years, ten thousand years, and a few centuries ago. Some common themes in all of these examples are that climate and life have been intertwined since the dawn of earth history, that relatively small causes have had large and often unexpected consequences, and that the magnitude of some of the possible global changes of the next century rival the magnitude of some of the biggest changes from our past.

Of course, the global changes of the past were not caused by humans. Nor could the animals of the earth, or the humans, do anything about these changes. Animals, plants, and humans moved, adapted, or died. In contrast, the global changes of the present and near future may be caused by humans, and perhaps we can do something about them.

The first example of global change is taken from more than a billion years ago. Early, innovative forms of life, blue-green algae, used the energy from the sun to split molecules of water and carbon dioxide and then recombined them differently to form organic compounds and oxygen--a process we call photosynthesis. Fossilized deposits from this period are called stromatolites; present-day structures that resemble stromatolites grow today in warm, shallow seas. With the burial of the organic matter produced by photosynthesis, and with other geochemical changes, the

50

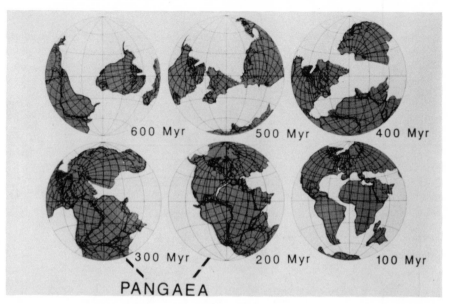

FIGURE 6.1 Locations of continents for various times during the past several hundred million years. (Adapted from N. Calder. 1983. Time-scale: An Atlas of the Fourth Dimension. Viking Press, New York.)

amount of oxygen in our ancient atmosphere began to increase and the amount of carbon dioxide began to decrease. High in the atmosphere oxygen was split apart by ultraviolet radiation from the sun and then recombined to form ozone. This marked the birth of our ozone shield. That shield is now about a billion years old. It has protected the earth's surface from harmful ultraviolet radiation and has permitted life to flourish on the continents ever since. It has been an old friend, and of course we need to know why it is becoming thinner now.

A second example of global change is taken from several hundred million years ago. The drifting and rifting of our restless continents created a grand sequence of global changes over hundreds of millions of years. Figure 6.1 shows the continents' locations from 600 million to 100 million years ago. During a particularly fascinating period between 300 million and 200 million years ago, the continents came together to form one supercontinent, called Pangaea, and then parted again.

We are using climate models, the same sort of models used to study possible future climates, to calculate the climate of an idealized Pangaean world (Figure 6.2). This one-continent world existed at the start of the age of the dinosaur. North America, Greenland, and parts of Eurasia were in the northern hemisphere. South America, Africa, India, Australia, and Antarctica were in the southern hemisphere. There was one ocean and one large sea, the Tethys Sea, on the eastern tropical shores. The huge land mass, according to the climate model, experienced huge seasonal swings of temperature. The continental interior was hot and dry, especially in summer, with temperatures exceeding 35°C, or 100°F.

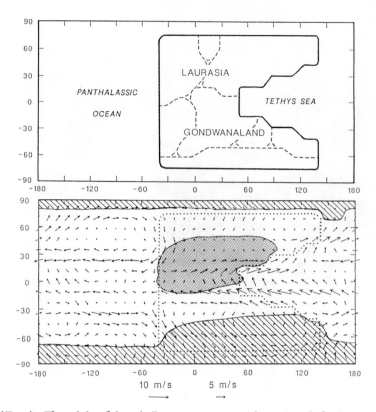

FIGURE 6.2 (Top) The idealized Pangaean continent with Laurasia in the
northern hemisphere and Gondwanaland in the southern hemisphere.
Panthalassa, the world ocean, and the Tethys Sea are indicated. Fine
dashed lines indicate very approximately the outlines of modern land
masses, but these outlines are only schematic. (Bottom) Surface winds
(arrows) and features of surface temperature (warmer than 30°C, stipple;
colder than 0°C, hatch) for June-July-August based on a climate model
simulation.

Polar regions were cold in winter, with temperatures below freezing. It
was humid along the coasts of the Tethys Sea, where monsoon winds were
strong.

Geologic evidence of the location of plant and animal fossils,
ancient sand dunes, and mineral deposits supports much of this computer-
estimated climate scenario. This agreement of the simulated climate with
geologic data is important. It shows that we are beginning to understand, and model, the processes of early global change. Although that
world was vastly different from ours in many respects, the climate model
calculates that the average temperature of the Pangaean world was only
about 5°C, or 9°F, higher than earth's present temperature. That is, the
Pangaean world was only marginally warmer than some projections of global
temperature for the next century.

Throughout this period of drifting continents there was a general
rise in the diversity and abundance of life on our planet. Figure 6.3

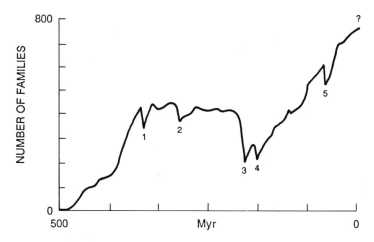

FIGURE 6.3 Over the past 500 million years, the general rise in the number of families of marine life has been interrupted at least five times by major extinctions. (Adapted from T. H. van Andel. 1985. <u>New Views on an Old Planet: Continental Drift and the History of Earth</u>, p. 282. Cambridge University Press, New York.)

illustrates the increasing number of families of marine life over the period from 500 million years ago to the present. However, the general rise has been interrupted at least five times by major extinctions of life. Around the time of Pangaea, the greatest extinction of all time occurred. By some estimates, about one-half of all families and three-fourths of all species became extinct. This great dying may have been related to the drastic changes in climate that accompanied the formation of Pangaea. Perhaps the Pangaean world had too few unique habitats, perhaps the heat and aridity were too extreme, perhaps ocean currents were different, or perhaps the amounts of oxygen and carbon dioxide were different. We do not know what caused the great extinction, but we know that it happened.

Can we learn, from such unexplained catastrophic extinctions, any lessons about the delicate balances that govern life on our planet? I hope so. Because some ecologists estimate that the destruction of tropical rain forests, along with other habitats elsewhere, may produce extinction rates over the next century that will rival the five great extinctions of the past.

Another great dying, the most recent, occurred around 65 million years ago. The dinosaurs died. Perhaps changes internal to our globe caused this event, too; or perhaps, as some have argued, an asteroid hit the earth. According to one theory, a giant dust cloud may have been thrown into the atmosphere by the impact. This cloud could have screened out the sunlight and caused a brief but deadly cold, dark winter over the face of the earth. For whatever reason, and there are many theories, the

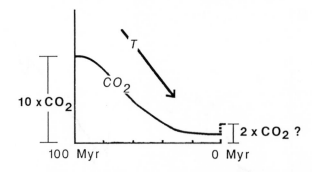

FIGURE 6.4 Over the past 100 million years, earth's climate has
experienced a long cooling trend, as indicated by the schematic arrow
labeled T for temperature. Over the same period, the atmospheric
concentration of carbon dioxide (CO_2) may also have declined from levels
10 times the present. For comparison, the estimated abrupt doubling of
CO_2 concentration is shown with a dashed line.

dinosaurs are gone! If this great dying had not occurred, dinosaurs
might still be the dominant species. But it did occur, and soon
thereafter, mammals became the dominant species.

A third example of global change encompasses a long slide from a warm
to a cool climate (Figure 6.4). It began around 100 million years ago,
when the climate was still much warmer than it is at present. Why was it
warmer? One strong possibility is that the carbon dioxide content of the
atmosphere may have been about 10 times the present level. This high
level could have been caused by great volcanic eruptions associated with
rapid spreading of the seafloor and rifting of the continents at that
time. However, as the continents moved toward their modern locations,
the rate of seafloor spreading slowed, and volcanic activity and
outgassing of carbon dioxide decreased. This, according to one theory,
caused the atmospheric concentration of carbon dioxide to fall. And as
the greenhouse effect diminished, the earth cooled. In other words, this
ancient global change may also have linked changes of carbon dioxide and
climate, but with falling levels of greenhouse gases and temperature. If
the amount of carbon dioxide doubles in the next century, it will
possibly mark a return to the higher carbon dioxide levels, and higher
temperature levels, of several million years ago. This is a graphic
illustration of the potential magnitude of the experiment that we
embarked on with the rapid burning of fossil fuels--fossil fuels that
were formed, in many cases, about 100 million years ago.

A fourth example of global change comes from the glacial cycles of
the past million years (Figure 6.5). The diminished greenhouse effect
mentioned above, perhaps aided by mountain uplift of the Rockies and in
Tibet, seemed to help set the stage for the growth of glaciers and huge
ice sheets. Geologic data for the most recent glacial age, which
occurred about 18,000 years ago, indicate that glacial ice covered much

DATA

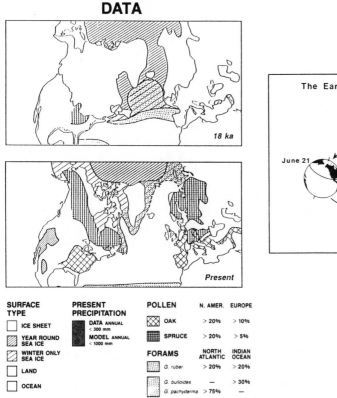

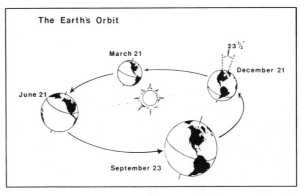

FIGURE 6.5 (Left) Changes in the earth's climate and vegetation that accompanied the transition from glacial conditions (18 ka, around 18,000 years ago) to interglacial conditions (present), as illustrated by geologic and paleoecologic evidence. (Right) Changes in the earth's orbit, shown here for modern conditions, are thought to pace the timing of glacials and interglacials. The minimum earth-sun distance occurs now in northern winter but cycles through the calendar year with a period of about 20,000 years. The axial tilt, now 23½°, varies between about 22° and 25° with a period of about 40,000 years. (Left: Reprinted, by permission, from COHMAP Members, 1988. Copyright (c) 1988 by the AAAS. Right: Reprinted, by permission, from N. G. Pisias and J. Imbrie. 1986/1987. Orbital geometry, CO_2, and Pleistocene climate, Oceanus 29(4):43. Copyright (c) 1986 by Woods Hole Oceanographic Institution.)

of North America and Europe and reached south to the locations of modern-day cities such as Chicago, New York, London, Berlin, and Moscow. For comparison, current maps show glacial ice only on Greenland.

In the past decade, great progress has been made in understanding the cause of these huge swings from glacial to nonglacial climates. Almost certainly, the swings are triggered, or paced, by relatively small changes in the earth's orbit, small changes that alter the amount of sunlight reaching the earth. These small changes in amount of sunlight equal only a few percent, but the consequences are large. What are these

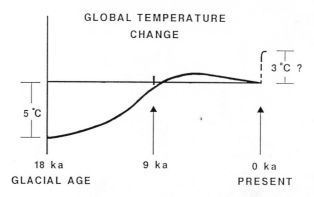

FIGURE 6.6 The simulated gradual increase in surface temperature,
averaged for the globe, from the glacial age (18,000 years ago) to the
present is about 5°C (9°F). For comparison, the estimated abrupt
increase in temperature of about 3°C over the next century due to
greenhouse warming is shown with a dashed line.

orbital changes? The earth's axis wobbles, like the axis of a spinning
top, completing one wobble cycle in about 20,000 years. Also, the tilt
of the earth's axis changes slightly, taking about 40,000 years for one
cycle. And the orbit itself alternates between being almost circular and
slightly egg-shaped, with a cycle time of about 100,000 years. When the
earth's orbit favors less sunlight in summer, the climate cools,
mile-high mountains of ice rise, and the sea level falls by several
hundred feet. When the orbit favors more sunlight in summer, the climate
warms, the ice melts, and the sea level rises.

What lessons can we learn from our glacial past? One lesson is that
small causes (such as small changes in the earth's orbit) may have large
consequences. There are also potential amplifiers in the global system.
In the case of glacial cycles, we now know that the amount of carbon
dioxide in the earth's atmosphere drops during the glacial swings, making
them even colder, and climbs as the ice melts. Thus changes in
greenhouse gases may amplify the consequences of a relatively small
initial change in the amount of sunlight.

Our roots, and our common past, have links to these glacial ages.
Some of our ancestors painted pictures of elk on the walls of the caves
of Southern France, sheltered from the cold northerly winds blowing
across the tundra of Ice Age Europe. Others took advantage of the
lowered sea level to cross the Bering Straits from the Old World to the
New World. Then, we humans could only react to nature's moves. Now we
are making the moves; we are causing global changes.

What happened when the recent glacial age ended? Starting about
18,000 years ago there was a gradual global warming trend of about 5°C,
or about 9°F (Figure 6.6). (Of course, much greater warming occurred in
polar regions.) The warming trend extended over more than 10,000
years--until roughly 6,000 years ago. That gradual warming trend is
contrasted with a projected, abrupt, 3°C warming trend in the next

DATA

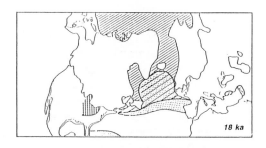

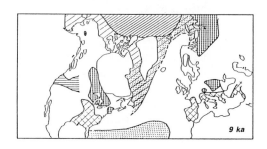

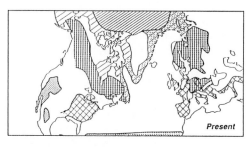

FIGURE 6.7 Changes in the earth's climate and vegetation that accompanied the transition from glacial conditions (18 ka, around 18,000 years ago) to rapid deglaciation (9 ka, around 9,000 years ago) to the present, as illustrated by geologic and paleoecologic evidence (see Figure 6.5 for key). (Reprinted, by permission, from COHMAP Members, 1988. Copyright (c) 1988 by the AAAS.)

century. This is a rather startling comparison. The temperature increase during the next century could be similar, in magnitude, to the entire temperature increase that has occurred since the last glacial age.

How have natural communities responded to these climatic changes of the past? One example is that whole forest communities have marched across the land. About 18,000 years ago, spruce forests were located in the central United States south of the ice sheet (Figure 6.7). As the ice melted and the climate warmed, the spruce forest moved to the Great Lakes area 9,000 years ago and is currently located in southern and northwestern Canada.

Some forests moved more than 500 miles to the north in the 10,000-year period of gradual warming, or about 5 miles per century. For comparison, the projected midlatitude warming trend in the next century might force our forests to try to march 250 miles per century--in other words about 50 times faster than the most recent natural rate. The inability of plant and animal communities to move together at such rates might literally tear communities apart. Even at the more sedate pace of the warming at the close of the last glacial age, there were extinctions--of the mammoth, for example, perhaps due to its failure to adjust to the changing environment, or to human impacts (such as hunting pressure), or to both; we do not know.

Climate in the tropics undergoes changes, too. Tombstone-like towers that are eroding fragments of lake sediments and that contain the skeletons of fish and crocodile bones are standing today in North Africa in the heart of the Sahara Desert. Around 5,000 to 10,000 years ago a vast lake covered the region, and a whole network of lakes and Neolithic fishermen occupied the Sahara. We have assembled geologic data to describe the climate of that time. Some 9,000 years ago, the area of wetter climate included much of North Africa, the Middle East, and parts of southern and eastern Asia (Figure 6.8). Experiments with climate models have shown that these huge changes in precipitation were also caused by small changes in the earth's orbit.

The important point is that, with the help of climate models, we now know how these changes came to be. About 9,000 years ago, the earth's orbit favored more summer sunlight. The land became hotter and the monsoon winds blew more strongly from sea to land. Rainfall increased and pushed farther inland. The increased rain created lakes in shallow depressions of the desert floor and, in parts of North Africa and the Middle East, created conditions more favorable for the agricultural revolution then under way. The fair agreement between model results and the geologic record of this global change gives us confidence that we are beginning to have a crude predictive capability for understanding what happened and why.

A fifth and last set of examples of global change comes from the present millennium. The golden age of the Mesa Verde Indians of Colorado may have been cut short by problems of overpopulation and overuse of the land. These problems were perhaps accentuated by persistent drought that began abruptly in the late thirteenth and early fourteenth centuries. The droughts in Mesa Verde began about the time that temperatures in Europe fell (Figure 6.9). An important point is that these relatively recent climatic changes can be dated very accurately. And from this we have learned that the climate can change abruptly.

Starting in the fourteenth century and continuing through the nineteenth century, Europe was colder than it is now (Figure 6.9). This period, called the Little Ice Age, has been illustrated by a nineteenth-century artist's sketch of the advance of a mountain glacier in Switzerland. In the nineteenth century, the glacier had descended to a valley floor, threatening a village. In the twentieth century, as a recent photograph shows, the glacier has retreated many miles up to the head of the valley. The climate has warmed from the nineteenth to the

DATA

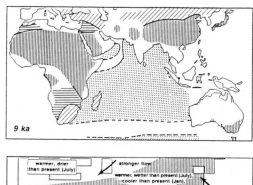

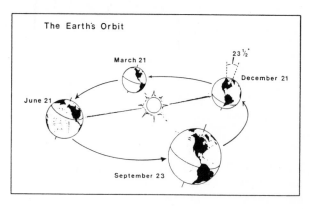

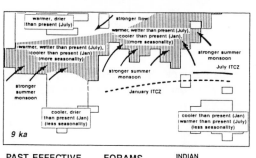

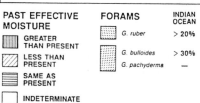

FIGURE 6.8 (Left) Features of the earth's climate around 9,000 years ago (9 ka) based on geologic and paleoecologic evidence (top panel) and climate model simulations of enhanced monsoonal circulations (bottom panel). (Right) Changes in the earth's orbit from the present configuration, where perihelion (minimum earth-sun distance) is in northern winter, to the configuration for 9,000 years ago, where perihelion was in northern summer and the axial tilt was 24°, account in climate model simulations for the enhanced monsoons. (Left: Reprinted, by permission, from COHMAP Members, 1988. Copyright (c) 1988 by the AAAS. Right: Reprinted, by permission, from N. G. Pisias and J. Imbrie. 1986/1987. Orbital geometry, CO_2, and Pleistocene climate, Oceanus 29(4):43. Copyright (c) 1986 by Woods Hole Oceanographic Institution.)

twentieth century, and present-day temperatures are already warmer than they were at any time in the last millennium.

We do not know, for certain, the causes of the climatic changes of recent centuries and decades. Perhaps there have been small changes in the amount of sunlight, or small changes in the frequency of volcanic eruptions, or subtle internal oscillations of atmosphere and ocean. These recent climatic changes are smaller in magnitude than the changes suggested by our models for the twenty-first century. Nevertheless, even these small changes are obviously large enough to have major regional

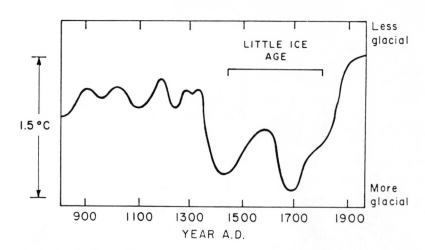

FIGURE 6.9 Estimates of the changes in temperature in Europe over the past 1,000 years. (Reprinted, by permission, from J. Imbrie and K. P. Imbrie. 1986. Ice Ages: Solving the Mystery, p. 181. Harvard University Press, Cambridge, Mass. Copyright (c) 1988 by John Imbrie and Katherine Palmer Imbrie.)

impacts, as indicated by the advance of glaciers in Switzerland of the thirteenth century or by the Dust Bowl years in America of the twentieth century. This backdrop of ongoing natural climate variability has another very serious consequence. It complicates our task of recognizing the initial phases of human-caused climate change--witness the discussions of the hot, dry summer of 1988.

To sum up, I have highlighted several examples of global change from the past: the dawn of life on our planet, the restless rifting of the continents, the waxing and waning of ice ages and monsoons, and the droughts and cold spells of recent centuries. There are a number of lessons to be learned, I think, from this brief historical perspective on global change throughout the millennia.

1. Climate and life have been intertwined since the dawn of earth's history.

2. Relatively small causes (such as orbital changes) have had large consequences.

3. Global change has sometimes been accompanied by the growth or catastrophic decline of species; only five great dyings in the past may compare in magnitude to some estimates of near-future extinctions from the diverse global changes now under way.

4. The potential magnitude of climatic change in the next century, caused by human activities, is comparable to that of some of the large natural climatic changes of the past, but human-caused changes may occur at much faster rates.

5. The global system we need to understand is complicated, but we are making progress in understanding how it works and in constructing predictive models.

Perhaps the most important lesson is that even though there is still much that we do not know about global change, past or future, we already know enough to begin to act now.

REFERENCES FOR ADDITIONAL READING

COHMAP Members. 1988. Climatic changes of the last 18,000 years: Observations and model simulations. Science 241:1043-1052.
Crowley, T.J. 1983. The geological record of climatic change. Rev. Geophys. Space Phys. 21:828-877.
Kutzbach, J.E., and R.G. Gallimore. 1989. Pangaean climates: Megamonsoons of the megacontinent. J. Geophys. Res. 94:3341-3358.
Schneider, S., and R. Londer. 1984. The Coevolution of Climate and Life. Sierra Club Books, San Francisco, 317 pp.

7

MATHEMATICAL MODELING OF GREENHOUSE WARMING: HOW MUCH DO WE KNOW?

J. D. Mahlman

For many decades scientists have known that a buildup of carbon dioxide (CO_2) in the atmosphere has the potential for warming earth's climate through the so-called "greenhouse" effect. Over the past 10 years, awareness has grown that other greenhouse gases can contribute in total to climate warming at a level comparable to that of CO_2. These include human-produced chlorofluorocarbons (CF_2Cl_2, $CFCl_3$, and others), methane (CH_4), and nitrous oxide (N_2O). The atmospheric concentrations of these gases are currently increasing at a rate sufficient to produce substantial atmospheric consequences over the next century.

These other greenhouse gases are well known to contribute to very significant expected changes in the atmospheric ozone structure and amount. Their potential to add to the CO_2 climate warming effect is not as universally appreciated. Here, I will emphasize only the expected climatic effects of the ensemble of greenhouse gases.

The information that I will present is derived from three-dimensional mathematical models of the climate system. A simplified schematic of the various relevant physical processes is given in Figure 7.1. Such comprehensive global climate models have been under intense development at the National Oceanic and Atmospheric Administration's (NOAA) Geophysical Fluid Dynamics Laboratory (GFDL) for over 25 years. Climate models have grown slowly but steadily in scope, complexity, and computational resolution over that period. Accompanying this growth is an improvement in the ability of the models to simulate the current climate. Accordingly, modeling atmospheric responses to changing conditions (e.g., seasonal and daily cycles, different planets (Mars and Venus), and ice age conditions) has become progressively more accurate.

Unfortunately, substantial uncertainties remain due to deficiencies in scientific understanding and insufficient computer power. However, significant progress is expected on both fronts over the next 10 years. Computer power is increasing while its relative cost is still decreasing. The impact of model dependence on computer power may be seen in Figure 7.2. Deficiencies in scientific understanding of such areas as ocean circulation, cloud processes, land surface processes, and chemical interaction will continue to yield gradually to intense scientific inquiry. In addition to the models, adequate data are necessary to evaluate results of the model calculations.

63

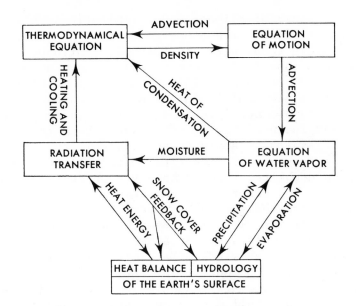

FIGURE 7.1 Simplified schematic diagram illustrating some of the interactive atmospheric processes governing earth's climate system. A proper climate model must account for all of these processes (and more) consistent with the laws of physics at a large number of points on the model "earth."

<u>Current Climate Model Resolution</u> (No Ocean)

5° latitude (300 n miles) X 5° longitude X 10 levels
(36 X 72 X 10 = 25,920 grid points)

<u>Improved Resolution</u>

2.5° latitude X 2.5° longitude X 20 levels
(207,360 grid points X <u>2</u>) (doubled time steps)
16 times the "cost"

<u>Exploratory Resolution</u>

1° latitude X 1° longitude X 40 levels
(2,592,000 grid points X <u>5</u>) (quintupled time steps)
500 times the "cost"

FIGURE 7.2 Illustrative example of the strong demands on computer resources as a climate model's grid resolution is increased.

In spite of significant climate model limitations, simulations of the effects of increases in the greenhouse gases permit a number of plausible inferences to be drawn today with considerable confidence. In most model studies to date, atmospheric CO_2 concentrations have simply been doubled (e.g., from 300 to 600 parts per million) and then maintained at that concentration until a new equilibrium climate is established in the model. Typically, such a model includes just the atmosphere; the only effect of the ocean included is its effect as a heat reservoir. We expect such model results to apply reasonably well to a combination of CO_2 and other trace gases where the total effect on the radiation budget is equivalent to a doubling of CO_2. Moreover, because of the now-recognized effect of the other greenhouse gases, calculations using doubling of CO_2 are now thought to be relevant (from a societal impact perspective) for conditions sometime into the middle of the next century. In this presentation, I will avoid detailed greenhouse gas scenarios. Rather, I will emphasize the kinds of expected impacts and my best estimates of their current scientific uncertainties.

Some of the possible climate responses to increased greenhouse gases are regarded to be rather well understood; others remain controversial. Scientific confidence is presented here in general terms. My estimates of confidence levels based on current models can be interpreted according to the following guidelines: By "virtually certain" I mean that there is nearly unanimous agreement within the scientific community that a given climatic effect will occur. Here, "very probable" means greater than about a 90 percent (9 out of 10) chance, and "probable" implies more than about a 67 percent (2 out of 3) chance. By "uncertain" in this context, I refer to an effect that has been hypothesized but for which there is a lack of appropriate modeling or observational evidence. I list below, in decreasing order of my current scientific confidence, some important model-predicted climate changes due to increased greenhouse gases. (This list is similar to that in National Research Council, 1987.)

o <u>Large stratospheric cooling (virtually certain)</u>. A reduction in upper stratospheric ozone by chlorine compounds will lead to reduced absorption of solar radiation and thus to less heating. Increased stratospheric concentrations of radiatively active trace gases will increase infrared radiative heat loss from the stratosphere. Decreased heating and increased cooling will lead to a marked lowering of upper stratospheric temperatures, perhaps by 10 to 20°C (Figure 7.3).

o <u>Global-mean surface warming (very probable)</u>. For a doubling of atmospheric CO_2 (or its radiative equivalent from all the greenhouse gases), the long-term global-mean surface warming is expected to be in the range of 1.5 to 4.5°C (see Figures 7.3 and 7.4). The most significant source of uncertainty arises from difficulties in modeling the feedback effects of clouds on climate change. The actual rate of warming over the next century will be governed by the growth rate of greenhouse gases, natural fluctuations in the climate system, and the detailed response of the slowly responding parts of the climate system, i.e., oceans and glacial ice.

o <u>Global-mean precipitation increase (very probable)</u>. As the climate warms, the rate of evaporation increases, leading to an increase

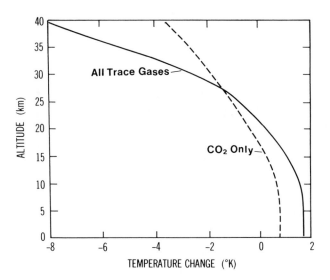

FIGURE 7.3 Estimate of the global-average temperature change (°C) for the year 2030 based on projected trace gas trends. "CO_2 Only" includes only effects of changing CO_2; "All Trace Gases" includes effects of changing CO_2 as well as all the other increasing greenhouse gases. (Reprinted from Ramanathan et al., 1987.)

in global-mean precipitation. Despite this increase in global-mean precipitation, local regions might well experience decreases in precipitation.

o <u>Northern polar winter surface warming (very probable)</u>. As the sea ice boundary is shifted poleward, the models predict a significantly enhanced surface warming in winter polar regions (see Figure 7.5). The greater fraction of open water and thinner sea ice is calculated to lead to an effective winter warming of northern polar surface air by more than 10°C relative to the current climate.

o <u>Reduction of sea ice (very probable)</u>. As the climate warms, total sea ice is expected to be reduced in response to warming in high latitudes of the Northern Hemisphere. However, new GFDL model results with a fully interactive ocean indicate little warming at Southern Hemisphere high latitudes over the next century, thus leading to little change in sea ice cover there (Figure 7.6). This new model result thus disagrees with the equilibrium model predictions for the Southern Hemisphere high latitudes as shown in Figure 7.5.

o <u>Northern high-latitude precipitation increase (probable)</u>. As the climate warms, the increased poleward penetration of warm, moist air may increase the annual-average precipitation and river runoff in Northern Hemisphere high latitudes.

o <u>Summer continental dryness/warming (probable)</u>. Several model studies have indicated a marked decrease of the soil moisture over some midlatitude interior continental regions during summer. This drying is

66

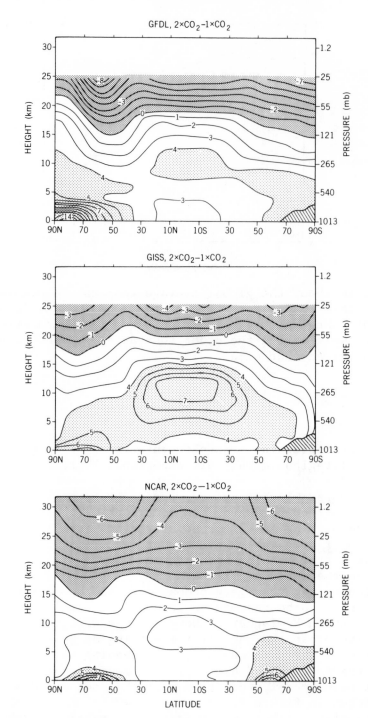

FIGURE 7.4 Latitude-height cross sections of December-January-February mean temperature change (°C) for a doubled CO_2 world compared to today's climate for 3 different models. Top, Geophysical Fluid Dynamics Laboratory (NOAA) model; center, Goddard Institute for Space Studies (NASA); bottom, National Center for Atmospheric Research. (Reprinted from Schlesinger and Mitchell, 1985.)

67

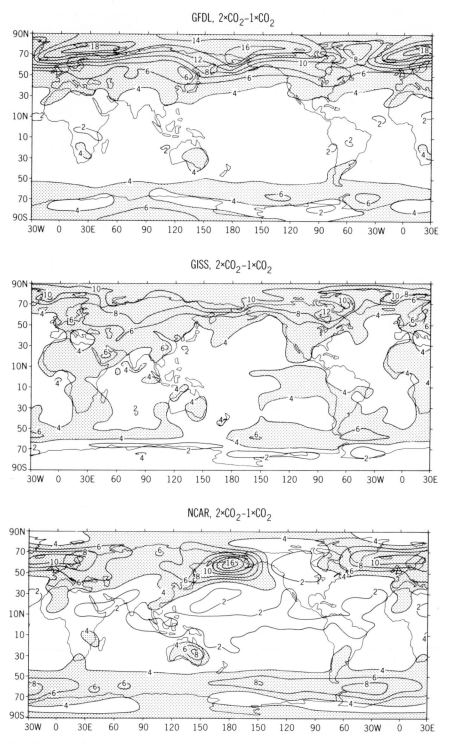

FIGURE 7.5 Latitude-longitude cross sections of December-January-February mean surface temperature change (°C) due to a doubled CO_2 as calculated by the three different climate models described in Figure 7.4. (Reprinted from Schlesinger and Mitchell, 1985.)

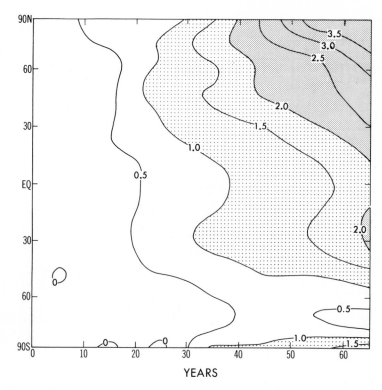

FIGURE 7.6 Time evolution of zonal-mean, decadally averaged temperature change (°C) in a GFDL/NOAA coupled atmosphere-ocean model due to a 1.0 percent per year buildup of CO_2. The resistance of the southern high-latitude region to greenhouse warming illustrates the potential of ocean circulation effects to yield results significantly different from those indicated in the previous generation of models (from Stouffer, R. J., S. Manabe, and K. Bryan, "On the Climate Change Induced by a Gradual Increase of Atmospheric Carbon Dioxide," submitted to Nature).

caused mainly by an earlier termination of snowmelt and rainy periods and thus an earlier onset of the normal spring-to-summer reduction of soil moisture. For a comparison of this effect in doubled and quadrupled CO_2 atmospheres, see Figure 7.7.

o Rise in global mean sea level (probable). A rise in mean sea level is generally expected due to thermal expansion of sea water in the warmer future climate. Far less certain are the contributions due to melting and calving of land ice. In addition, for the next century there now exists the possibility of increased snow accumulation over the antarctic continent. Predictions of actual changes in mean sea level thus remain difficult and controversial.

o Regional vegetation changes (uncertain). Climatic changes in temperature and precipitation of the kinds indicated above must inevitably lead to long-term changes in the surface vegetative cover. The exact nature of such changes and how they might feed back to the climate remain uncertain.

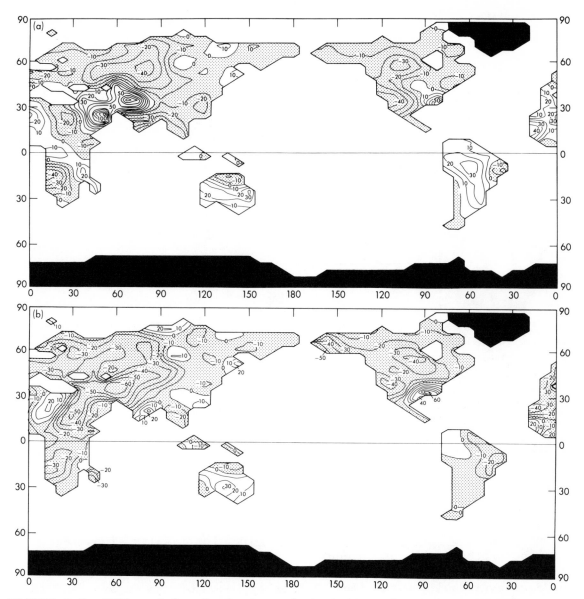

FIGURE 7.7 GFDL model calculation of changed soil moisture (%) for months of June-July-August. Upper picture is for doubled CO_2; lower picture is for quadrupled CO_2. Darkest shading indicates decreases greater than 20 percent. (Reprinted by permission from Manabe and Wetherald, 1987. Copyright (c) by the American Meteorological Society.)

o <u>Tropical storm increases (uncertain)</u>. A number of scientists have suggested that a warmer, wetter atmosphere could lead to an increased number and intensity of tropical storms, such as hurricanes. However, tropical storms also are governed by other factors such as local wind structure. At the present time, this effect has not been satisfactorily addressed in the coarse-resolution climate models due to the relatively small size of tropical disturbances.

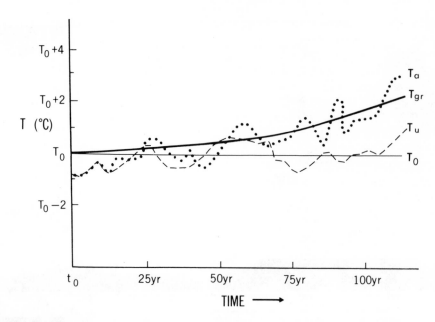

FIGURE 7.8 Schematic illustration of the greenhouse warming detection problem. The thick line (T_{gr}) represents evolution of a hypothetical average gradual greenhouse temperature warming. The thin line (T_0) is an assumed undisturbed time-averaged temperature. The dashed line (T_u) represents the actual temperature variation in an undisturbed climate. The dotted line (T_a) is the actual fluctuating signal for an earth with gradually increasing greenhouse gases. Note that it can take many years to separate the fluctuating greenhouse signal (T_a) from the undisturbed fluctuating signal (T_u).

o <u>Details of next 25 years (uncertain)</u>. The results given above describe expected changes in equilibrium climate due to hypothetical large changes in greenhouse gases. In actuality, these gases are increasing gradually with time. Initially, much of the excess heat is absorbed into the oceans, perhaps in complex ways we do not yet understand well. Further, we can expect that natural, decadal-scale climatic fluctuations due to interactions between the atmosphere and the oceans will continue to occur. The midwestern drought in the 1930s and the high water levels of the Great Lakes in the 1980s are good examples of such climatic fluctuations. On these shorter time scales, such natural fluctuations would artificially reduce or enhance the apparent greenhouse warming signals (Figure 7.8). Until such decadal-scale fluctuations are understood or are predictable, it will remain difficult to diagnose the specific signals of permanent climate change as they evolve over the next quarter-century. Moreover, detecting climate change signals becomes even more difficult when smaller regions and/or shorter periods of time are considered.

Even though the above uncertainties are daunting, important advances have already been achieved in the observation, understanding, and modeling of the climate system. The current models are capable of simulating the gross features of geographical and seasonal variations of the global climate. Furthermore, some of these models have achieved successful simulations of the very cold climate of the last glacial maximum and of the extreme temperatures found on the planets. These overall scientific advances have initiated the current public awareness of climate change and its potential implications for the future of the world. This awareness has escalated the need for reliable climate predictions, accurate assessments of the causes of the actual changes occurring, and an ability to distinguish human-produced climate change from longer-period natural variations. Although progress has been made, as noted above, significant deficiencies remain in the capability of the scientific community to address these needs.

Much more effort must be expended worldwide toward providing a climate monitoring and measuring system characterized by careful instrument calibrations and intercomparisons and a commitment to continue measurements over many decades. Focused research into climate processes must be accelerated so that theories can be formulated and reevaluated in the light of newer information. To reduce climate modeling uncertainty, it is imperative that climate modeling efforts receive state-of-the-art supercomputing resources. Additionally, new scientific talent must be developed to exploit those resources.

Through careful, long-term research on observation, modeling, and analysis, our scientific uncertainties will decrease and our confidence for predicting details of the climate system and its changes will gradually improve. A final, very confident prediction is that the societal need for accurate and detailed climate predictions will increase as fast or faster than the scientific community can provide them. The effort to meet these societal challenges will require the combined forces of the world scientific community in a sustained effort spanning decades.

ACKNOWLEDGMENTS

I would like to thank Dr. Syukuro Manabe and his group at the Geophysical Fluid Dynamics Laboratory/NOAA for their contributions to the perspectives offered here as well as for their unrelenting commitment to solving the greenhouse gas modeling problem. I am also indebted for the contributions and cooperative efforts from the other climate modeling groups at the National Center for Atmospheric Research, Goddard Institute for Space Studies/NASA, Oregon State University, and the United Kingdom Meteorological Office.

REFERENCES

Manabe, S., and R. T. Wetherald, 1987. Large-scale changes of soil wetness induced by an increase in atmospheric carbon dioxide. Journal of Atmospheric Science, 44, 1211-1235.

72

National Research Council, 1987. Current Issues in Atmospheric Change.
    National Academy Press, Washington, D.C.
Ramanathan, V., L. Callis, R. Cess, J. Hansen, I. Isaksen, W. Kuhn, A.
    Lacis, F. Luther, J. Mahlman, R. Reck, and M. Schlesinger, 1987.
    Climate-chemical interactions and effects of changing atmospheric
    trace gases. Reviews of Geophysics, 25, 1441-1482.
Schlesinger, M. E., and J. F. B. Mitchell, 1985. Model Projections of
    the Equilibrium Climatic Responses to Increased Carbon Dioxide.
    Projecting the Climatic Effects of Increasing Carbon Dioxide.
    DOE/ER-0237, U. S. Department of Energy, 80-141.

8

THE EARTH'S FRAGILE OZONE SHIELD

Susan Solomon

HISTORY OF THE OZONE DEPLETION PROBLEM

Ozone is an essential part of the earth's ecological balance because it absorbs certain wavelengths of biologically damaging ultraviolet light that are not effectively absorbed by any other component of the earth's atmosphere. The degree of protection provided by the ozone layer is related to the total amount of ozone between the sun and the planet surface, and hence to the total integrated column abundance (the total ozone). It is believed that the evolution of biological life on the planet surface was closely tied to the evolution of the protective ozone layer. Most the world's ozone is found in the stratosphere, at altitudes from about 10 to 35 km.

The study of atmospheric ozone and concern about its possible depletion dates back only to about the 1970s. During the middle and late 1970s, it was recognized that continued use of man-made chlorofluorocarbons could significantly perturb the natural ozone abundance.[1] Chlorofluorocarbons are used in a wide variety of industrial applications, including refrigeration, air conditioning, foam blowing, and cleaning of electronics components. Theoretical studies of the chemistry of ozone carried out in the year prior to the discovery of the antarctic ozone hole suggested that chlorofluorocarbon production would be expected to decrease ozone by perhaps 5 to 10 percent sometime in the next century.[2]

In 1985, scientists from the British Antarctic Survey reported observations of a 50 percent decrease in total ozone during the antarctic spring.[3] Figure 8.1 illustrates some of the observational data that revealed the ozone hole. This unexpected seasonal decrease in contemporary antarctic ozone was quickly dubbed the "antarctic ozone hole," and it rapidly captured worldwide attention. Laboratory, field, and theoretical studies over the past 4 years since the discovery of the antarctic ozone hole have led to a progressively clearer picture of how it takes place, why it takes place largely in Antarctica, and its likely implications for other latitudes. These investigations have changed the understanding of atmospheric ozone chemistry and have led to a heightened awareness of the importance of global change.

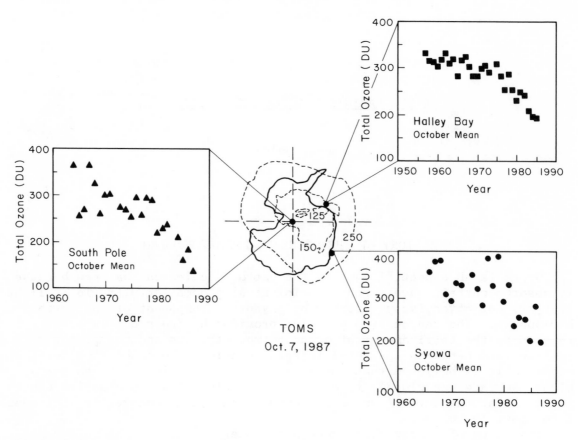

FIGURE 8.1 Total ozone. Observational data that first indicated the existence of the antarctic ozone hole. DU, Dobson units; TOMS, total ozone mapping spectrometer.

CURRENT THEORETICAL UNDERSTANDING OF ANTARCTIC OZONE DEPLETION

The key to antarctic ozone depletion is the extreme cold temperatures that occur in the antarctic stratosphere. The stratosphere is extremely dry, generally precluding significant cloud formation except under the coldest conditions. The occurrence of clouds changes the chemistry in a very fundamental way: it allows reactions to occur on surfaces rather than between gas molecules.[4] Chemical reactions take place on these surfaces, converting chlorine from forms that do not react with ozone to other, less stable forms that readily break up in the presence of sunlight and go on to destroy ozone. Both cold temperatures and sunlight are critical to the ozone depletion process. Therefore, antarctic ozone depletion does not take place during the winter, when temperatures are coldest but when the polar regions are largely in darkness, but rather in the spring, after sunlight returns and temperatures remain cold.

OBSERVATIONAL EVIDENCE

When the antarctic ozone hole was first discovered, little was known about the antarctic stratosphere beyond the ozone measurements themselves. There were virtually no available data on other chemical compounds present in the stratosphere, and there was also a pressing need for more detailed meteorological information. These needs were rapidly addressed by ground-based and aircraft expeditions to the Antarctic, during which state-of-the-art instrumentation was used to measure chemical compounds, to probe the nature of the polar clouds, and to further understand the meteorology.

Observations of a broad range of atmospheric compounds, including chlorine monoxide, chlorine dioxide, hydrochloric and nitric acid, nitrogen oxide and dioxide, and nitrous oxide, were rapidly obtained. The observations all display a highly unusual chemistry, greatly perturbed by the presence of clouds. The observed abundances of chlorine and bromine monoxide will result in rapid and substantial ozone loss similar to that observed in the antarctic spring. The chlorine monoxide levels found in Antarctica are of particular importance, since this species participates in catalytic cycles that rapidly destroy ozone. The abundances of chlorine monoxide have been shown to be about 100 times greater than expected in the absence of cloud chemistry. The broad range of experimental techniques used and the consistency of the observed perturbations in many different chemical compounds have provided firm evidence that these perturbations account for much if not all of the antarctic ozone loss.[5]

METEOROLOGICAL PROCESSES: ANTARCTIC AND ARCTIC

The study of atmospheric chemistry is highly interdisciplinary, with strong links to meteorology and radiative transfer. Meteorology plays an important role in setting the stage for polar chemistry and modulating the extent of ozone depletion. For example, some antarctic winters are warmer than others and are likely to exhibit fewer polar stratospheric clouds and less ozone depletion. Warmer winters are also likely to modulate the ozone abundances through direct meteorological effects. Meteorological processes also play a critical role in determining whether or not the depletion of polar ozone can spread to lower latitudes through mixing and large-scale overturning of the atmosphere.

There are a number of important differences between the antarctic and arctic stratospheres. Satellite and ground-based observations show ozone losses of about 5 to 10 percent in the arctic winter at high latitudes.[6] Clearly, the ozone depletion in the arctic stratosphere is thus far much smaller than that of the antarctic stratosphere. This is partly due to the fact that winter arctic temperatures are warmer on average than those of the Antarctic. Perhaps more importantly, the arctic stratosphere generally warms up much earlier in the spring season than does the antarctic stratosphere. This likely leads to a critical difference in the temporal overlap between the cold temperatures and the sunlight required for ozone depletion. It is critical to understand how the temperature history interacts with chemical processes and to evaluate whether an unusually cold and late arctic spring would result in substantial ozone losses there.

In Antarctica, the ozone loss of perhaps 50 percent is accompanied by a much more spectacular increase in chlorine monoxide by a factor of 100. The latter perturbation is much more readily identified as compared to natural variability, and implies that measurements of chemical species such as chlorine monoxide can help to evaluate the present and future potential for ozone loss in those environments where direct identi-fication of small ozone losses may be difficult. These considerations motivated studies of the chemical composition of the arctic stratosphere during the winters of 1987 and 1988, in which researchers sought to understand the chemistry of the Arctic during winter and to determine the extent to which it too may be influenced by polar stratospheric clouds.

IMPLICATIONS

Many scientists view the antarctic ozone hole as a sort of global early warning system. The unusual chemistry of polar stratospheric clouds has clearly made the antarctic ozone layer more vulnerable to anthropogenic chlorine than the rest of the contemporary atmosphere. An area of increasing concern is the possibility of similar chemical reactions occurring on the type of particles present at warmer latitudes, especially following major volcanic eruptions, which can greatly enhance the particles present in the stratosphere around the world.

It is clearly fortunate that the ozone hole has so far occurred largely in that part of the globe that contains the least biological life. Ongoing research is, however, aimed at studying the possible effects of ozone depletion on phytoplankton and, by extension, other creatures such as krill, penguins, and seals. It is of paramount importance to determine the origin of the smaller ozone changes measured at other latitudes and to evaluate the future changes that can be expected worldwide if mankind continues the emission of chlorofluoro-carbons.

NOTES

1. Molina, M. J., and F. S. Rowland, Nature, 249, 810, 1974; Stolarski, R. S., and R. J. Cicerone, Can. J. Chem., 52, 1610, 1974; a recent review has been given in McElroy, M. B., and R. J. Salawitch, Science, 243, 763, 1989.
2. National Research Council, Causes and Effects of Changes in Stratospheric Ozone: Update 1983, National Academy Press, Washington, D.C., 1984.
3. Farman, J. C., B. G. Gardiner, and J. D. Shanklin, Nature, 315, 207, 1985.
4. Solomon, S., R. R. Garcia, F. S. Rowland, D. J. Wuebbles, Nature, 321, 755, 1986; McElroy, M. B., R. J. Salawitch, S. C. Wofsy, J. A. Logan, Nature, 321, 759, 1986; Toon, O. B., P. Hamill, R. P. Turco,

77

J. Pinto, Geophys. Res. Lett., 13, 1308, 1986; McElroy, M. B., R. J. Salawitch, S. C. Wofsy, Geophys. Res. Lett., 13, 1296, 1986; Crutzen, P. J., and F. Arnold, Nature, 324, 651, 1986; Molina, L. T., and M. J. Molina, J. Phys. Chem., 91, 433, 1986; Molina, M. J., T. L. Tso, L. T. Molina, F. C. Y. Wang, Science, 238, 1253, 1987; Tolbert, M. A., M. J. Rossi, R. Malhotra, D. M. Golden, Science, 238, 1258, 1987.

5. deZafra, R. L., M. Jaramillo, A. Parrish, P. Solomon, B. Connor, J. Barrett, Nature, 328, 408, 1987; Brune, W. H., J. G. Anderson, K. R. Chan, submitted to J. Geophys. Res., 1989; Solomon, S., G. H. Mount, R. W. Sanders, A. L. Schmeltekopf, J. Geophys. Res., 92, 8329, 1987; Farmer, C. B., G. C. Toon, P. W. Shaper, J. F. Blavier, L. L. Lowes, Nature, 329, 126, 1987; the status of antarctic ozone research prior to August 1987 was reviewed in Solomon, S., Rev. Geophys., 26, 13, 1988; important new findings from airborne experiments will appear shortly in a special issue of J. Geophys. Res., 1989.

6. NASA reference publication 1208, Present State of Knowledge of the Upper Atmosphere 1988: An Assessment Report, National Aeronautics and Space Administration, Washington, D.C., 1988.

9

TERRESTRIAL ECOSYSTEMS

Peter M. Vitousek

INTRODUCTION

Terrestrial ecosystems are metabolic systems whose activity produces
and consumes many of the gases that drive global change. Plants use
nitrogen from the soil to capture energy from the sun and carbon dioxide
from the atmosphere. Soil microorganisms ultimately utilize much of that
energy, in the process releasing carbon dioxide and methane as end
products and nitrogen-containing trace gases as by-products of their
activity.

The amounts involved can be very large; terrestrial plants take up
more than 100 PG (billion metric tons) of carbon annually, and plants and
microorganisms return approximately as much to the atmosphere in respiration. This exchange is 20 times greater than the amount of carbon released by fossil fuel combustion. Similarly, fluxes of both methane and
nitrous oxide from terrestrial ecosystems are well in excess of fossil
fuel sources (Mooney et al., 1987).

The large absolute amount of material exchanged by terrestrial ecosystems does not mean that such systems control ongoing changes in the
composition of the atmosphere and hydrosphere. Gases released by terrestrial ecosystems may be more or less balanced by uptake in those systems (as for carbon dioxide), or they may be balanced by natural processes in the atmosphere (as for methane and nitrous oxide). However,
terrestrial ecosystems are capable of driving change when their own
dynamics are altered by human activity or by climate change. They are
equally capable of responding to global changes in ways that feed back
(positively or negatively) to those changes.

I will develop three points in this brief presentation. First,
although terrestrial ecosystems appear stable in the absence of human
intervention, they are in fact dynamic in ways that interact strongly
with the atmosphere and ocean. Second, human activity is now changing
the earth system in wholly novel ways and at rates far in excess of
any in the past several million years. These changes not only alter
terrestrial ecosystems directly, but they also interfere with the
capacity of those systems to respond to change. Third, our ability to
understand the workings of the earth system may at last be developing
as fast as our ability to alter the earth unintentionally. In that
there is hope.

TERRESTRIAL ECOSYSTEM DYNAMICS

Terrestrial ecosystems vary more or less repeatably on temporal scales ranging from days to hundreds of millennia. Some of the short-term changes, including seasonal and interannual variation in the absorption of photosynthetically active radiation, can be imaged directly by satellite sensors. When summed globally, these seasonal changes correlate very strongly with seasonal changes in carbon dioxide concentrations (Fung et al., 1987)--an exciting demonstration that satellite-based earth observation now allows truly global-scale research. Repeated observations over periods of years and decades will allow us to determine the effects of occasional events such as El Nino or the 1988 drought and of directional changes in the distribution of vegetation. However, despite the value of such time-series measurements and the great interest in global change, they are very difficult to support--as is indicated by the ongoing drama over the continuation of LANDSAT.

On the other extreme, terrestrial ecosystems vary on time scales of tens of thousands to hundreds of thousands of years as a consequence of glacial-interglacial cycles. "Ice ages" have been a cyclical feature of the earth for millions of years; the cause of these regular cycles is variations in the earth's orbit (Hays et al., 1976; Imbrie et al., 1984), although not all of the mechanisms between orbital cause and climatic effect have been worked out in full.

Compared to the present, full-glacial periods were (obviously) much cooler, especially at higher latitudes (CLIMAP Project, 1976). The sea level was lower due to accumulation of more of the earth's water in ice, and circulation patterns of the oceans differed substantially. There were also correlated patterns of reduced carbon dioxide (and methane) concentrations in the atmosphere (Barnola et al., 1987), and the consequent reduction in the greenhouse effect contributed to the overall cooling of the earth.

These full-glacial conditions altered terrestrial ecosystems spectacularly. The major vegetation zones were often shifted thousands of kilometers from their present positions, the fraction of the earth's surface covered by different types of vegetation was altered substantially, and many ecosystems were composed of combinations of species wholly different from those found anywhere today (Davis, 1981). The co-occurrence of reduced vegetation cover and higher wind speeds caused greatly increased wind erosion at times during glacial cycles, leading to the deposition of large amounts of terrestrially derived nutrients into the sea.

Changes in the composition of the atmosphere could also have had direct effects on terrestrial ecosystems. There are two great photosynthetic pathways in land plants, termed the C_3 and C_4 pathways for the number of carbon atoms in the first organic product of carbon dioxide fixation (Bjorkman and Berry, 1973). The C_4 pathway, found primarily though not exclusively in tropical grasses, actively concentrates carbon dioxide within leaves. Its activity is therefore less sensitive to external carbon dioxide concentrations than is that of the C_3 pathway (Strain and Bazzaz, 1983). Low concentrations of carbon dioxide in the full-glacial atmosphere should therefore have favored C_4-dominated

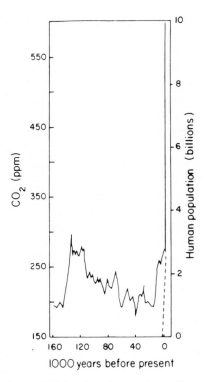

FIGURE 9.1 Past and projected variations in the concentration of carbon
dioxide (solid line). The 160,000-year record is derived from the
Vostok ice core (Barnola et al., 1987); the modern values are measured
(to 350 ppm) or projected. Carbon dioxide in the atmosphere has already
increased nearly as much (in 200 years) as the entire range of the
160,000-year record. The dashed line is the human population--past and
projected.

ecosystems such as tropical savannas over C_3-dominated ecosystems such as
tropical forests. These two represent sharply defined alternative states
in many tropical regions today; they differ strikingly in their carbon
storage, albedo, effect on the local climate, and fire regime. A substantial expansion in savanna caused by reduced CO_2 in the atmosphere
could therefore feed back to climate and the composition of the atmosphere.

MODERN GLOBAL CHANGE AND TERRESTRIAL ECOSYSTEMS

How does human-caused change compare with past changes such as the
glacial-interglacial cycle? At least in terms of the composition of the
atmosphere, the ongoing change is both much larger and much faster
(Figure 9.1). The concentration of carbon dioxide varied from 200 to
285 ppm during the glacial-interglacial cycle; it is now approximately
350 ppm, and it is increasing rapidly. This increase will accentuate the
greenhouse effect, the more so because greenhouse gases other than CO_2

are also increasing rapidly as a consequence of human activity. The climatic effects are predicted to be similar in distribution but opposite in direction to those during full-glacial periods; temperatures will likely increase substantially at high latitudes and relatively little at low latitudes.

In addition, all else being equal, the increase in carbon dioxide concentration should favor C_3-dominated forest vegetation over C_4-dominated savanna. The net result would be an increase in carbon storage on land and therefore a buffering of the rate of increase in the atmosphere. However, all else is decidedly not equal. Humans are clearing and burning tropical forests at unprecedented rates. Much of the land so cleared is converted to cattle pastures through planting of C_4 grasses, either immediately upon clearing or after 1 or 2 years of cropping. This activity can itself increase the amount of carbon dioxide in the atmosphere-ocean system. It also changes local, and possibly regional, climate because pastures (like savannas) have higher albedo, higher surface temperatures, and lower near-surface humidities than do the forests from which they are derived.

Tropical deforestation has many other effects. It increases transport of dissolved and particulate nutrients to water systems. Biomass burning adds nitric oxide to the remote atmosphere, where it catalyzes the production of tropospheric ozone--and during the dry season ozone concentrations in the Amazon and Zaire basins are approaching the intolerable regional levels of eastern North American and northern Europe (Browell et al., 1988). (They are still far from those in southern California.) Nitrous oxide produced in tropical pastures may be a significant source of global increase in that greenhouse gas, and cattle themselves are a globally significant source of methane. (Tropical and subtropical rice paddies--themselves derived by land conversion--are the most important source of methane worldwide (Cicerone and Oremland, 1988)). Further, deforestation is causing the extinction of numerous species--a global change that is significant in its own right, and one that forecloses forever any possibility of the reconstitution of tropical forests as they exist today.

I have concentrated on changes in the tropics, but changes to and within terrestrial ecosystems in other areas may be equally significant globally (Schimel et al., 1989). Of particular concern are the following:

1. The effects of increased carbon dioxide concentrations on the functioning of terrestrial ecosystems everywhere. Increased concentrations are known to affect plant growth, water use efficiency, nutrient use, decomposition, and herbivory under controlled conditions; their long-term interactive effects on the ecosystem level are worth exploring.

2. The possibility that global warming will catalyze the release of vast amounts of organic carbon stored in high-latitude soils. To the extent that this takes place in wetlands, an increase in methane in the atmosphere will result; to the extent that it occurs in upland sites, carbon dioxide concentrations in the atmosphere-ocean system will increase.

3. The possibility that atmospheric transport and deposition of nitrogen-containing compounds resulting from human activities (fossil fuel combustion, fertilizer use) will alter the metabolism of extensive areas of temperate forests and grasslands downwind of industrial and agricultural areas. This process ultimately may alter the amounts and kinds of compounds exchanged between terrestrial ecosystems and the atmosphere or hydrosphere.

CONCLUSIONS

These are exciting times--the ability to understand many aspects of the earth system is within our reach for the first time, and public, educational, and scientific interest in global change is overwhelming. However, the global changes that have taken place to date are small relative to what can be expected in the next 50 years. Unless surprising progress is made, carbon dioxide concentrations soon will be more than 50 percent greater than the preindustrial values; most tropical forests worldwide will be a memory. Scientific conclusions, partial though they inevitably will be, must be transformed into global action with unprecedented speed if our increased ability to understand the earth is to hold out any hope against the exponential increase in human-caused global change.

REFERENCES

Barnola, J.M., D. Raynaud, Y.S. Korotkevich, and C. Lorius. 1987. Vostok ice core provides 160,000-year record of atmospheric CO_2. Nature 329: 408-414.

Bjorkman, O., and J. Berry. 1973. High-efficiency photosynthesis. Scientific American 229 (Oct): 80-93.

Browell, E.V., G.L. Gregory, R.C. Harriss, and V.W.J.H. Kirchoff. 1988. Tropospheric ozone and aerosol distributions across the Amazon Basin. Journal of Geophysical Research 93: 1431-1451.

Cicerone, R.J., and R.S. Oremland. 1988. Biogeochemical aspects of atmospheric methane. Global Biogeochemical Cycles 2: 299-327.

CLIMAP Project. 1976. The surface of the ice-age earth. Science 191: 1131-1137.

Davis, M.B. 1981. Quaternary history and the stability of forest communities. Pp. 132-153 in West, D.C., H.H. Shugart, and D.B. Botkin (eds.), Forest Succession. Springer-Verlag, New York.

Fung, I.Y., C.J. Tucker, and K.C. Prentice. 1987. Application of Advanced Very High Resolution Radiometer vegetation index to study atmosphere-biosphere exchange of CO_2. Journal of Geophysical Research 92: 2999-3015.

Hays, J.D., J. Imbrie, and N.J. Shackleton. 1976. Variations in the earth's orbit: pacemaker of the ice ages. Science 194: 1121-1132.

Imbrie, J., J.D. Hays, D.G. Martinson, A. McIntyre, A.C. Mix, J.J. Morley, N.G. Pisias, W.L. Prell, and N.J. Shackleton. 1984. The orbital theory of Pleistocene climate: support from a revised

83

chronology of the marine 180 record. Pp. 269-305 in Berger, A.L. (ed.), Milankovitch and Climate. D. Reidel Publishers, Dordrecht, The Netherlands.

Mooney, H.A., P.M. Vitousek, and P.A. Matson. 1987. Exchange of materials between terrestrial ecosystems and the atmosphere. Science 238: 926-932.

Schimel, D.S., M.O. Andreae, D. Fowler, I. Galbally, R.C. Harriss, H. Rodhe, B. Svensson, and G. Zavarzin. 1989. Key areas for research in global trace gases exchange. In Andreae, M.O., and D.S. Schimel (eds.), Exchange of Trace Gases Between Terrestrial Ecosystems and the Atmosphere. John Wiley and Sons, Chichester, in press.

Strain, B.R., and F.A. Bazzaz (chairmen). 1983. Terrestrial plant communities. Pp. 177-222 in Lemon, E.R. (ed.), CO_2 and Plants: The Response of Plants to Rising Levels of Atmospheric Carbon Dioxide. Westview Press, Boulder, Colorado.

10

HUMAN DIMENSIONS OF GLOBAL ENVIRONMENTAL CHANGE*

Roberta Balstad Miller

We have known for centuries that man has the power to alter the surface of the earth. A thousand years ago an anonymous Anglo-Saxon poet described the ox that pulled his plow through newly cleared fields as the "grey enemy of the wood." But although the poet could see the gradual transformation of forests into farmland, he could not see the far-reaching second- and third-order consequences of the deforestation of the British Isles. Today we know that the changes set in motion by the ancient plowman did not stop with the destruction of the woods but led to widespread alterations in the land itself, affecting soil drainage, erosion, and fertility and eventually altering the climate that surrounded his descendants.

WHAT HAVE WE LEARNED?

In the last 100 years, we have learned a great deal more than the Anglo-Saxon poet could have guessed about the variety of ways that we human beings continuously transform the earth. We have learned, first, that seemingly insignificant patterns of behavior, repeated over long periods of time, can have major consequences for the environment. For example, we have seen that such basic agricultural activities as plowing and grazing have caused and continue to cause radical changes in the landscape that can lead to severe environmental problems. We have learned, further, that the soil erosion resulting from agriculture may be a far more serious source of water pollution than industrial effluents, that wildlife and the continued survival of certain plant species are as threatened by agricultural expansion as by the more visible encroachment of urbanization, and that land use patterns may be as critical a factor in climate change as are the burning of fossil fuels and the emission of chemicals by modern industry.

_____

*This paper was published in a somewhat different form as "Global Change Research Challenges Social Sciences" in The AAAS Observor, July 7, 1989, p. 5. Copyright (c) 1989 by the American Association for the Advancement of Science.

Second, we have learned that modern technologies, both through their demands for energy and through the life-styles they foster, are one of the major sources of global environmental change. At the present time, moreover, technology-based environmental change is intensifying. The rapid advances in technology characteristic of the twentieth century combined with the dissemination of these technologies across the globe conspire to increase anthropogenic or human influences on the environment.

What is particularly chilling about this source of environmental stress is that we have not had enough experience with these technologies or with their by-products to know what their long-term effects will be. We know that small manufactories from the mid-nineteenth century continue to pollute running water more than 100 years later. But the long-term influence of the chemical and technological residues of our own day--the insecticides and fertilizers, plastics and detergents, vehicular and industrial emissions--are generally unknown to us. We do know that 100 years from now, they will play a more significant role in shaping the environment of our grandchildren than the abandoned mills of the nine-teenth century have played in shaping our environment.

Third, we have learned that one of the major threats to the earth's environment is the sheer mass of the earth's population. For example, one of the most important predictors of carbon dioxide production in any area is population density. Even if the population of the earth were stable, this would be a cause for concern. But our population is not stable; it is increasing more rapidly at the present time than it has at any previous point in the earth's history. Equally important, this growth in the world population is accompanied by an increase in the per capita consumption of goods. Together, the growth in population and the growth in consumption continually accelerate the complexity and magnitude of human impacts on the environment.

WHAT IS BEING DONE?

At the present time, the scientific community is engaged in a major international effort to study, model, and predict changes on the surface of the earth and in the atmosphere above the earth. Scientific attention has focused, however, on physical and biological processes of change. The activities of the Anglo-Saxon plowman and his successors are not seen as an integral part of research on these physical and biological pro-cesses.

Instead, the human dimension has been isolated from ongoing scientific research on the earth system and relegated to distinctly separate spheres called "social science" and "policy." It has been isolated not because it is unimportant, but because of the argument that the complexity of examining biogeochemical changes on a global scale will not permit the addition of so messy a set of analytic variables as humans and the institutions they create.

This situation is changing. Increasingly, physical scientists are recognizing that their knowledge of physical processes of terrestrial or atmospheric change is incomplete without some understanding of the human dimensions of this change or of the ways that human action sets physical processes in motion and modifies ongoing processes. Similarly, biologists and ecologists realize that a critical element in their study of ecological systems is human action. They have concluded that for scientific reasons they can no longer ignore human interactions with the environment.

THE NATURE OF SOCIAL SCIENCE RESEARCH ON GLOBAL CHANGE

Parallel to this growing recognition among natural scientists of the need for research on the interactions of human, biological, and physical systems in global change is a similar interest within the social science community. Social scientists argue, however, that the nature of the research that is needed is broader than the natural scientists realize. Because natural scientists are primarily interested in physical and biological change, they recognize the need to learn more about anthropogenic influences on such change, that is, the direct interaction between human behavior and physical systems. But social scientists argue that to understand the human dimensions of global change, we must also understand patterns of behavior and interactions far more complex than that relatively straightforward nexus between the individual and the environment.

There are two major reasons for this expansion of the scope of research. First, research on the human dimensions of global change must deal with a changing target over time, one that encompasses both human action and human reaction to the environment. We disturb the universe, to echo T. S. Eliot, and then we change our behavior in environmentally significant ways--in response both to the environmental changes we initiate and to other factors totally unrelated to the environment.

A second reason that the human dimensions of global change are far more complex than the physical or natural dimensions of such change is that human action is embedded in institutions and cultures. We are influenced by economic, cultural, and political forces as well as by individual motivations. Research on the human dimensions of global change that ignores institutional imperatives, that ignores the various economic and political influences on people in different nations, that ignores the cultural diversity that distinguishes and, in some respects, dominates our actions, would be nearly as inadequate as research that ignores the human dimension altogether.

THE RESEARCH AGENDA

What is needed, and what is currently being started in a number of countries, is a series of broad social science research programs on the human dimensions of environmental change. Such research programs should be able to accomplish two tasks: (1) They should feed into the physical and natural science research activities already under way, and (2) they should simultaneously be concerned with those elements of the global change research agenda that are purely social and economic but that ultimately are as powerful determinants of environmental change as on-going physical and biological processes. This dual research agenda is rather daunting, yet both aspects are necessary.

In addition, social science research on the human dimensions of global change must be global in scope, international in organization, institutional in focus, and historical in breadth. It must be global in scope because we are dealing with processes of change and interactions that are global. At the present time, social science research focuses on units of analysis that fall on a continuum stretching from the case study to the national study. Comparative cross-national research is an exception to this general practice, but even in comparative studies, research most frequently consists of a comparison of national phenomena or behavior in two or more nations.

To understand global change, however, we must go beyond the nation state in defining research topics, for it is clear that environmentally significant events and actions work together across national boundaries. This is not to negate national influences or activities. National regulations and laws, for example, continue to be important influences on global change both in the nation and in larger geographic areas. However, if we are to understand the very critical interaction between the nation state and global change, national phenomena must be examined within the larger global context. A similar interaction takes place in local and regional environmental change. Whatever the geographic scale of our research, we must deal with the interactions between small-scale and global phenomena.

The research program must also be international in organization. This is not the same as global. Global refers to space; international refers to political units across the globe. If there is to be a successful social science research program on global environmental change, it must involve collaboration and cooperation among social scientists of many nations. This is necessary to foster standardization of data collection, to test hypotheses in various settings, and to understand the role of cultural influences in shaping environmental attitudes.

Social science research on global change must be institutional in focus. Environmentally significant human action is determined by structural and institutional requirements and by limitations ranging from national laws and regulations to profit margins, transportation patterns, agricultural markets, and tax structures, to name a few. To ignore the structural and institutional influences on human behavior is to close one of the most important windows we have on understanding environmental change.

Finally, this research must be historical. By this, I mean it must be concerned with human and institutional activities over long periods of time. Because time constitutes an active variable in social behavior, social scientists tend to define the time limits of their research very carefully and, in practice, very narrowly. In empirical research projects in particular, the time period covered by social science research is generally quite short. But the study of global change requires research over long time periods, and if we are to understand human interactions with the environment in their full complexity, social scientists must be prepared to undertake research projects that examine changes over decades, or even centuries.

RESEARCH AND ENVIRONMENTAL POLICY

If such a research program were in place in the United States and in other nations, would it tell us what to do about the environment or how to reduce the adverse effects of human action? Unfortunately, it would not. There is a rapidly growing appreciation of the environmental problems caused by human action, and we recognize increasingly that we must respond to these problems as individuals and as nations before the changes we are setting in motion cause irreversible damage to the environment. There is a sense of urgency about the problem that is quite appropriate, given its magnitude.

Yet we must be cautious about looking to research, whether in the social or the natural sciences, for definitive answers to difficult policy questions. Research is essential to understand the processes of change in the environment, and that understanding is critical for the ability of governments to make wise policy choices. Sound environmental policy must be _informed_ by scientifically sound research in the social and natural sciences. But research in the social sciences or the natural sciences cannot--and should not--be used to prescribe a course of action for governments to take.

This is not to underestimate the need for social science research on the human dimensions of global change. But it is to emphasize that the role of social science research in public policy is one of illumination, of informing the policy process. Goethe's last words on his deathbed were "_mehr licht_" (more light). I think they describe quite well what social science contributes to environmental policy--more light, essential light, without which our policies will be inadequate to meet the challenge of our deteriorating environment. More light, without which we will never be able to comprehend the complex interrelationships between human beings and the globe we inhabit.

SUMMARY

In conclusion, the following points summarize important aspects about the study of human systems in global change. First, cumulative and intensifying environmental change on a global scale is the major problem facing the world at this point in our history. The scientific information we need must encompass physical, biological, and social processes of change.

Second, in examining human and social forces in environmental change, we must deal both with direct human action and indirect institutional and structural causes of change in the earth system. We must examine deforestation and the market economy that makes it profitable. We must examine ozone depletion and the regulatory climate that fosters continued use of chlorofluorocarbons. The topic of indirect structural and institutional influences on environmental change is one that requires a great deal of fundamental research in the next several years.

Third, scientific research is essential to understand the environmental processes that are transforming the earth, but research itself cannot determine public policy. Sound national and international policies for coping with environmental change must be informed and clarified by scientific research, and particularly by social science research, but ultimately environmental policy, if it is to be effective, must be determined through the political process. In most nations, environmental policies will be produced within the framework of a social and political consensus on what constitutes just and responsible action. This is a task that will increasingly absorb the energies of our governments over the next several decades, and if we expect to have wise and effective policies, we need to ensure that government officials have access to the best and most complete scientific understanding of environmental change.

Writing in the eighth century, Bede compared the life of man on earth to the flight of a sparrow through a hall in which people are eating and talking on a winter's day. Outside it is cold and stormy. Inside it is warm and light. The sparrow flies in through one door and quickly out the other. While he is inside, he is safe from the storm, but after this brief moment of comfort he vanishes into the dark winter. "In the same way," Bede wrote, "the life of man on earth is only a short space. Of what went before and what follows after we know nothing."

In some sense, this analogy applies to our study of global change. We have, until this point, concentrated on what is lighted and comfortable, that is, the observable, measurable processes of change in concrete physical and biological systems. The next step must be to examine the darkness, those unknown processes of change operating outside the hall, the difficult analytical task of understanding the human dimensions in environmental change. If we do not venture outside the lighted area, we will have a fair chance of understanding what is going on within eyesight, but we will be condemned to having incomplete knowledge of the world. However, if we go beyond this, outside the comfort of what we know through empirical observation of physical systems, if we dare to challenge the darkness, we will be frequently confused, at times overwhelmed, but ultimately on the right track to understanding global change.

11

THE HUMAN CAUSES OF GLOBAL ENVIRONMENTAL CHANGE

B. L. Turner II

Recognition of humankind as an awesome power in transforming the earth is not new. George Perkins Marsh heralded this theme in the nineteenth century, W. I. Vernadsky, among others, recognized the impending global dimensions of this power in the first half of this century, the now classic work <u>Man's Role in Changing the Face of the Earth</u>, (Vol. 1 and 2, William L. Thomas, Jr., ed., University of Chicago Press, Chicago, 1971) placed it in historical context, and recently the sustainability of human uses of the biosphere has been explored. The human impress on the earth--its present magnitude and enormous potential for still further transformation--is now conventional wisdom. This wisdom is complemented by the maturation of the study of the human causes of change. We have identified many of those human actions that are the major sources of environmental change, and we continue to document the degree to and manner in which they propel the transformation of the biosphere. In concert with these efforts, the rudiments of a conceptual framework that groups and links the components and processes of human-induced change have emerged. That the general structure of this framework has remained relatively unchanged during this half-century implies a consensus about it. Major research efforts are now aimed at refining the framework by detailing the relative weights and positions of its components and, more recently, by grounding it within a general understanding of human behavior.

The conceptual framework we seek is one that will help us understand the three principal types of environmental change: those in material and energy flows, in biota, and in the physical structure of the biosphere (Figure 11.1). It must be remembered, however, that while the relationships between these changes and human activity are the ultimate focus of study, these relationships are mediated by the very nature of the physical environment in which they are taking place:' a tropical wetland, a midlatitude desert, or a semitropical grassland, for example. In our quest to understand the global condition, we must not lose sight of the fact that human actions are very much grounded in place, that the differences among places are extreme, and that the human action and environmental change relationship will vary accordingly. The human side of the framework is characterized by at least three broad parts: the proximate sources of change, the human driving and mitigating forces, and human behavior.[1]

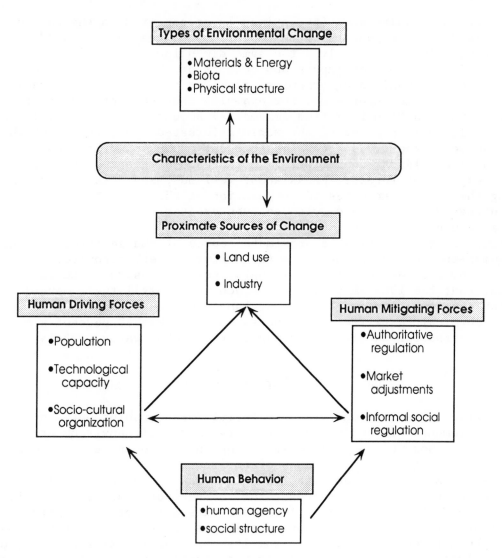

FIGURE 11.1 Components and linkages of the human causes of global environmental change.

PROXIMATE SOURCES OF CHANGE

The proximate sources of change constitute the near-end or end products of human activity whose immediate consequences are alterations and transformations of the environment. As such, proximate sources can be viewed as lenses through which human forces are directly translated into environmental change; they represent, therefore, the observational points of departure for empirical examinations of the human dimensions of global change.

A simple scheme recognizes two major proximate sources: those of land use and industry.[2] Land use involves the kind, scale, and spatial distribution of human activity, primarily as captured in the changes made in the landscape components of the environment, the so-called "faces of the earth." Industry involves the same dimensions, but primarily as captured in the flows of inputs and outputs of the production processes that affect the basic biogeochemical flows of the biosphere. Examination of land use change informs us, for example, that the net loss of forests owing to human action from time immemorial amounts to about 8 million km^2 or about 15 to 20 percent of the world's forested area; that over the last 300 years there has been a 450 percent expansion of the world area devoted to croplands, constituting an increase of about 12.4 million km^2, at the expense of forests, grasslands, wetlands, and deserts; and that during the past 3 centuries the population of the largest 40 urban areas has increased some 25-fold, with an even greater estimated increase in land area under urbanization.[3]

Studies of industry as a proximate source of change inform us that the human-induced contribution to the sulfur budget, accounted for almost entirely by the burning of fossil fuels, now exceeds the natural contribution; that the 1980 emissions of carbon from fossil fuels were 3 times greater than those from all forms of land use change; and that the global use of chemical fertilizers has increased more than 83 percent since 1950, exacerbating nutrient loadings in the global water systems.

HUMAN DRIVING AND MITIGATING FORCES OF CHANGE

Empirical examinations of the material relationships between the various elements of land use and industry, and their environmental impacts are, however, only the first step in developing a more complete understanding of the human causes of environmental change. These relationships must be linked with the underlying or deeper social forces that drive the proximate sources. For the sake of clarity, it is important to distinguish between two broad categories of these forces. The first category involves those attributes and actions of humankind that have altered the biosphere from its state as it existed at the dawn of civilization, before human impact was significant, particularly at the global scale. These I refer to as the human driving forces of change. In the second category are those actions that, both indirectly and directly, impede, alter, or counteract the driving forces or their impacts. These actions that diminish the net change are referred to as human mitigating forces.

In reality this division is not exclusive. Mitigating forces can have unintended consequences such that they themselves become driving forces. For instance, flood control dikes built in China over the centuries had the unintended consequence of raising riverbeds, which created even larger problems of flooding. And the switch in the Basin of Mexico to unleaded gasoline, undertaken to reduce lead emissions to the sensitive atmosphere, has greatly increased the production of ozone, creating major new impacts on local vegetation. Mitigating forces sometimes divert the driving forces of environmental change rather than preventing or eliminating them and, in these cases, play a role that is complementary to driving forces. The appearance of new driving forces of change may also mitigate other changes, as when the adoption of a new energy source relieves the impacts created by the source it replaced. Our discussion is simplified, however, if the two types of forces are distinguished.

Human Driving Forces

Human driving forces are the sum of individual and group actions, but more manageable, collective categories of these actions or action and impact relationships are needed for comparative analysis of the type that is especially important for understanding global change. Only broad typologies or classifications of these forces have so far been developed. Population, or more correctly, population change, constitutes one category that is almost universally recognized and, as noted below, is logically and factually fundamental among the driving forces. There is less agreement over the other possible categories and their relative positions within our framework. Here, we recognize technological capacity, a category that has received considerable attention, and sociocultural organization, which encompasses a myriad of proposed forces, including culture, institutions, and social, economic, and political structure.

The strong impact of population on the transformation of the earth is undeniable and well known and does not need to be expounded at length here. Population is a fundamental driving force because each individual minimally requires living space, shelter, food, and water, regardless of the social and technical actions that determine how these needs are met. The demand for these basic needs stimulated by population growth over the last 300 years, but especially during the past 50 years, has created exceptional levels of "living pressures" on that 25 percent of the world that historically has been the most intensely utilized, and it has led to new levels of pressures in other areas, most of which are thought to offer more constraining conditions for human use.[4]

The amount of and trends in global transformation attributable to the basic needs of population are difficult to specify for a number of reasons that cannot be detailed here, such as the variable meaning of "basic" for different cultures and socioeconomic conditions at different times. We can, however, draw on case studies for illustrations and insights. The ninth-century Classic lowland Maya civilization serves as an interesting baseline for comparison with the present. It involved a vibrant and healthy population in a premodern but sociotechnically advanced culture that met its basic or subsistence needs but probably did not consume excessively beyond them.[5] The population, situated in the heartland of the civilization at the base of the Yucatán Peninsula, grew for over 2 millennia, peaking at densities that may have exceeded 100 people per square kilometer. The environmental impacts were considerable, particularly in the transformation of the landscape. Ultimately, perhaps more than 75 percent of the upland tropical forest was altered for agriculture, settlement, and fuel, and 10 percent of the wetlands was transformed for cultivation through ditching and mounding, resulting, at a conservative estimate, in 0.4 ha of land altered per capita for cultivation and in 0.5 ha of total land altered during times of peak population. These changes created an "open landscape" and exacerbated problems of soil erosion and nutrient sequestering, both of which may have had major impacts on ground water in this karstic zone. I suggest that not more than 10 percent of this change was attributable to a surplus of consumption and production over the requirements for comfortable subsistence. This transformation of the heartland was driven largely by population growth, although the causes for that growth are speculative at best.

But of course, as technology has developed, along with the organizational capacity of society to employ it, so has its role as a driving force in amplifying and extending the range of the impacts of population alone. Technological capacity is considered a driving force because it can amplify the very demand for resources and because its development has both intended and unintended consequences for the environment.

At least two qualities of technology have been fundamental to its role as a driving force; these are elaboration and mobilization. The first involves the development and expansion of technological knowledge; the second refers to the organizational ability to marshal the materials and energy required for the use of that knowledge. The premodern Maya, for illustration, mobilized local human labor and materials to transform primarily regional landscapes through land clearance, creating paved civic and ceremonial centers, terraces and field walls, and wetland fields and canals.

Modern world history stands in stark contrast to the Maya case. The elaboration of technological capacity is such that enormous, global-scale increases in production capacity have been possible, even at lowered energy efficiency: witness the global increase in food production, which has kept pace with global population growth, in part through technologies that have increased the intensity of production by some 80 percent since the midpoint of this century. This elaboration through fossil fuel technologies has increased enormously the impacts on the basic biogeochemical flows of the biosphere, placing this kind of transformation on a par with or ahead of basic landscape change. The spatial (and temporal) separation of production and consumption has increased: witness the almost 20-fold rise in the total tonnage of international seaborne freight during this century and the 7-fold per capita increase, even accounting for the massive growth in global population. And the spatial separation of the sources and consequences of environmental change is also increasing: witness the afforestation of many regions and zones within the developed world, despite the overall unprecedented density of settlement there, and witness also the location of the so-called "ozone hole" in relation to the locations of the major activities that have created it.

The third and most diffuse category of driving forces is sociocultural organization, the intricate web of economic and political structures, and social values and norms. This category is critical to global environmental change in several ways. The forces within it are instrumental in driving the level of demand for physical resources well beyond basic subsistence needs, in the extreme case creating a culture of mass consumption such that items or levels of consumption once considered luxuries are seen as necessities. The estimated surplus of 10 percent over subsistence needs in Maya production and consumption pales in comparison to modern conditions throughout the developed world. For example, various estimates indicate that, accounting for waste and loss, the average American consumes between 3400 and 3600 kcal of food each day, perhaps 50 percent more than that required for basic needs.[6] Associated with these same forces has been the "democratization of accumulation," which has made possible the satisfaction of that increased demand by giving rise to the capacity for mass consumption throughout the wealthier segments of the world.[7] The impacts apparently extend even to the most socially planned societies of the developed world; for instance, not only has Sweden's consumption of material goods increased by more than 200 percent since 1950, but much of it is also apparently related to the desire to keep up with stylistic changes, such as in kitchenware.[8] Of course, this phenomenon is not new; even the Maya built larger and more elaborate palaces and pyramids as their wealth permitted, but the range and magnitude of this accumulation were severely restricted. Whereas we estimated 0.5 ha per capita of the total land altered by the Maya, the current global average for cultivated and grazing land alone is about 1 ha per capita.

It is noteworthy that these changes in the forces of demand and accumulation affect the distribution of resources and of the associated environmental impacts. The inequities that exist among regional patterns of consumption are well known. For instance, the wealthy countries of the world, constituting about one-quarter of the global population, consume 80 percent of the world's commercial energy; stated another way, each person in the "have" world consumes, on average, about the equivalent of 32 barrels of crude oil per year, whereas each person in the "have not" world consumes only 3.5 barrels. Likewise, the sources of environmental consequences, many global in scale, are unequal; for example, about 40 percent of all carbon dioxide emissions to the atmosphere, apparently leading to global climatic change, is attributable to seven wealthy countries, comprising only 11 percent of the world's population.[9]

Other sociocultural variables may significantly influence the changes made in the environment. This subject requires scrutiny because little effort has been expended so far toward empirical assessments of it. One variable cited as important is environmental ethics and attitudes. As an example, premodern cultures are commonly taken to have shown greater care for the environment than do modern ones. Yet examples of environmental degradation among these cultures exist, be it salinization of irrigated lands in the Tigris-Euphrates basin or massive deforestation in the central Maya lowlands. Whether these were exceptions to a general rule and whether they represent mismanagment that had clear-cut cultural and population consequences remain to be documented.

It has been claimed that political organization of resources affects the degree of environmental impact. Yet it appears that many centrally planned economies with state control over physical resources have problems of environmental degradation similar to those of capitalist economies based on private ownership, although pure cases of each system are difficult to find. Common or public property is often identified as either unusually prone or unusually resistant to environmental degradation, but in fact, it seems to vary greatly in that regard according to circumstances.[10] In the absence of sufficient comparative assessments, these roles of culture remain in question; the degree of influence they may have beyond population, technology, and wealth has not been established.

Human Mitigating Forces

To the driving forces of human-induced change and their environmental consequences are counterposed the human mitigating forces. It is perhaps important to recognize that the most straightforward force of this kind, direct regulation of the exploitation of specific resources, is not a recent phenomenon. For example, concerns about deforestation in sixteenth-century Britain, seventeenth-century France, and eighteenth-century Russia led to regulations designed to control and monitor forest loss. But, of course, international, national, and local institutions dedicated to mitigating environmental change have proliferated in the latter half of this century.

Mitigating forces can be viewed in terms of the particular driving forces that they counter, distinguishing, for example, those that slow population growth from those that decrease technological waste output. An alternative approach identifies the means by which the mitigation is effected. Based on this approach, three types of mitigating forces can be identified: those based on authoritative regulation, on market adjustments, and on informal social regulation. Authoritative regulation involves the enactment of rules enforced by a designated control agency that can take punitive action against violators. Market adjustments refer to changes in production and consumption as influenced by changes in economic value and cost, for example the relationship between gasoline or water consumption and its cost to the consumer. Informal social regulation includes at least two broad types: social norms and values that give rise to shared views or practices, and changes in production and consumption based on reasons other than purely economic ones. Of the three types, informal social regulation is the least understood, and its mitigating impacts are the most contentious among students of the subject.

Those forces of mitigation involving regulatory environmental control have been most successful when addressed to specific environmental issues and driving forces or to specific regions and zones, for instance, the regulation of the whaling industry, the soil conservation measures implemented on the Great Plains of North America after the Dust Bowl, the decrease in the global use of DDT, and the apparent leveling off of industrial emissions in much of the developed world. It is important, though, to distinguish mitigation effected because of environmental concern from that resulting from other processes: oil price increases in the 1970s and consequent market adjustments, rather than authoritative regulation, were responsible for a significant part of the reduction in automotive pollution at that time. It is also important to recognize that mitigating forces apparently have had their greatest success within the well-developed service and industrial economies of the world, indicating the significance of wealth in the larger structure of mitigation.[11]

Identification of the key driving and mitigating forces and documentation of the manner in which they interact to generate the proximate sources of change--land use and industry--are essential contributions of the social sciences to our current study of global environmental change. Attention must be given to the ways in which these forces are affected by changes in scales of time, space, magnitude, and clustering. Just as agglomeration of economic activities increases production efficiency, so too may the agglomeration of production and consumption lead to changes in the driving and mitigating forces, in the proximate sources, and, ultimately, in the environment. For example, some evidence and interpretations indicate that the increasing size of cities is related to increasing per capita municipal waste.[12] And a recent attempt to examine global historical trends in the changes in some of the components of the biosphere suggests that even with expanded production and consumption, mitigating forces designed to counteract some industrial effluents have led to decreases in the rates of delivery of some chemical elements and compounds to the environment. It is note-worthy that the impacts of such mitigation were realized quickly, after only some 10 to 20 years of regulation, in part because the major sources of the effluents are found within those regions that have the capacity to enforce regulation.

HUMAN BEHAVIOR

Future research on driving and mitigating forces obviously must be anchored on a strong, empirical base that documents the relationships among them and with environmental change. Complementing this research, however, must be an increase in the efforts to place these forces in the contexts of individuals and of society at large. To understand why the human forces operate as they do--that is, to explain why humankind transforms the biosphere in the way it does--ultimately will require us to embed these forces and the changes that they produce within an under-standing of human behavior. This will require that the frontiers of the study of global environmental change be expanded and that the natural sciences work hand-in-hand with the social sciences and the humanities.

This objective, of course, is complex and cannot be detailed here. Theories of behavior have been proposed from various realms of knowledge, and each realm claims privileged status in the explanation of human behavior. This poses no small problem for which resolution must be sought. Research may indeed contribute to a resolution by examining human behavior on several planes. Clearly much is to be learned from detailed assessments of behavior as it occurs in specific contexts. Likewise, much is to be gained from the search for broader, pan-cultural attributes of human behavior that can be linked to environmental change. As we contemplate the impact of humans on nature, we need to understand our humanity and the unique ways in which we, as a unique species, use and alter the earth. Problems to explore include the variable roles of human territorial strategies and the conditions that give rise to this

98

variability; the ability to expand both the number of niches used and the
ways in which they are used; the persistence of cultures in the face of
global interconnectivity; the changing array of and flexibility in the
pathways by which societies draw sustenance from nature; and the will-
ingness and ability of humans to alter their behavior reflexively when
confronted with new conditions or information.

The heritage of social science research on human relationships with
nature and on human agency in environmental change has begun to unravel
the complex networks of the linkages involved. This research informs us
of both the awesome power of human actions to transform the earth bene-
ficially and detrimentally, and of the ingenuity of humankind in adjusting to its environmental blunders, albeit not without substantial and
unequal sacrifices.[13] The messages from this work and its brief exploration here are simple. We are the cause and the solution, and both of
these attributes of ourselves are extremely complex and are made even
more so by their connection to the global environmental system.

NOTES

1. Human driving and mitigating forces constitute the core of all
    frameworks. These are typically linked to the proximate sources of
    change or to human behavior, depending on the primary objective of
    the study. Here all three parts are included to illustrate the
    relationships among them.
2. See, for example, W. C. Clark, The Human Dimensions of Global
    Environmental Change, report prepared for the National Research
    Council's Committee on Global Change, in Toward an Understanding of
    Global Change: Initial Priorities for the International Geosphere-
    Biosphere Program, National Research Council, 1988.
3. Unless stated otherwise, estimates of global change are taken or
    derived from the papers of the 1987 Earth Transformed symposium held
    at Clark University. Revised papers are forthcoming in B. L. Turner
    II et al., eds., The Earth as Transformed by Human Action, Cambridge
    University Press, Cambridge, 1990.
4. See, for example, P. A. Sanchez and S. W. Buol, Soils of the Tropics
    and the World Food Crisis, Science 188, 1975. Very few examples
    exist of good quality agricultural soils that are sparsely occupied
    or utilized. One such is the central Maya lowlands of the lower
    Yucatán peninsular region.
5. The Classic lowland Maya represent a much more sophisticated case
    than, say, hunter-gatherers in that the range of resource use was
    larger and involved a small portion of luxury production and con-
    sumption. The assessment presented is culled from an extensive
    literature, but see B. L. Turner II, Issues Related to Subsistence
    and Environment Among the Ancient Maya, in Prehistoric Lowland Maya
    Environment and Subsistence Economy, M. Pohl, ed., Peabody Museum,
    Harvard University, Cambridge, Mass., 1985, pp. 195-209; B. L. Turner
    II, The Rise and Fall of Population and Agriculture in the Central
    Maya Lowlands: 300 BC to Present, in Hunger in History: Food

Shortage, Poverty and Deprivation, L. Newman et al., eds., Basil Blackwell, London, in press.

6. Most of the energy and food estimates have been taken or derived from V. Smil, Energy, Food, Environment: Realities, Myths, Opinions, Clarendon Press, Oxford, 1987.

7. The term "democratization of accumulation" was suggested to me by R. W. Kates.

8. L. Uusitalo, Environmental Impacts of Consumption Patterns, St. Martin's Press, New York, 1986, pp. 78 and 104.

9. For carbon dioxide, Communications, Carbon Dioxide Information Center (Oak Ridge National Laboratory), Winter 1989; for energy, United Nations Conference on Trade and Development, Handbook of International Trade and Development Statistics, 1981 Supplement, UNCTAD, New York, 1982.

10. B. M. McCay and J. M. Acheson, eds., The Question of the Commons: The Culture and Ecology of Communal Resources, University of Arizona Press, Tucson, 1987.

11. See R. W. Kates, B. L. Turner II, and W. C. Clark, The Great Transformation, in The Earth as Transformed by Human Action, B. L. Turner II et al., eds., Cambridge University Press, Cambridge, 1990, forthcoming.

12. From Uusitalo, 1986 (above), p. 74.

13. I am grateful for discussions with and comments from R. W. Kates, W. C. Clark, W. B. Meyer, J. Emel, R. C. Mitchell, and the various members of the Graduate School of Geography, Clark University, associated with The Earth as Transformed by Human Action program. The paper, of course, is my responsibility.

PART C   IMPACTS OF GLOBAL CHANGE

WHAT DOES GLOBAL CHANGE MEAN FOR SOCIETY?

Lester R. Brown

To summarize what the changes in the earth's natural systems and resources mean for society and to establish a framework for further discussion, I will draw heavily from the Worldwatch Institute's State of the
World reports, the annual assessments we launched in 1984.[1]
    The key question is, How will the changes in the earth's natural
systems and resources affect us? We know that we cannot continue to
damage our life-support systems without eventually paying a heavy price.
But how will we be affected? What will the price be? Is it likely to be
a buildup of carcinogens in the environment so severe that it increases
the incidence of cancer, dramatically raising death rates? Or will the
rising concentration of greenhouse gases make some regions of the planet
so hot that they become uninhabitable, forcing massive human migrations?
Will depletion of the ozone layer in the upper stratosphere that protects
us from ultraviolet radiation lead to serious health problems--a rising
incidence of skin cancer, eye damage, including earlier cataract formation, and the suppression of human immune systems? Or could it be
rising sea levels on a scale that would force a relocation of hundreds of
millions who live only a few feet above current sea level? Or will it be
something that we cannot now even anticipate?
    The answer is that it will probably be all of these, although some
will affect us more and sooner than others. We know from geographic
analyses of epidemiological data that there are cancer hotspots in some
industrial regions, but the increased number of cancer cases in the
United States that is attributable to toxic chemicals is still minuscule
compared with the 390,000 deaths per year attributed by the surgeon
general to cigarette smoking. The chemical age is still young; it may
take time for the full health effects of exposure to toxic chemicals to
become evident.
    If the buildup in greenhouse gases continues and temperatures rise as
projected, some regions could become uninhabitable. A glimpse of such a
future was seen last summer in the Yangtze Valley of Central China, where
temperatures rose above 100°F for several consecutive days. One

---

[1]I am indebted to my colleague, John Young, for his assistance with
research and analysis and to Sandra Postel and Christopher Flavin for
their review and constructive suggestions.

manifestation of heat stress was the dramatic rise in heatstroke victims. In the central China cities of Shanghai, Nanjing, and Wuhan, hundreds died and local hospitals were overrun with heatstroke victims.

When will the consequences of the global changes that we have set in motion be felt? Will we pay the price or will our children? Or, as with the irreversible loss of biological diversity, will it be all generations to come? Some of the global changes, such as rising sea level and ozone depletion, are not likely to have serious social consequences until we are at least a few decades further down the road. Others may cause dislocation in the next few years.

Amidst all the uncertainty, food scarcity is emerging as the most profound and immediate consequence of global change, one that is already affecting the welfare of hundreds of millions of people. All the principal changes in the earth's physical condition--shrinking forests, deteriorating rangelands, soil erosion, desert expansion, acid rain, stratospheric ozone depletion, the buildup of greenhouse gases, the loss of biological diversity, and the dwindling per capita supplies of cropland and fresh water--are having a negative effect on the food prospect.

The first concrete economic indication of broad-based environmental deterioration now seems likely to be rising food prices. They have the potential to disrupt economies and, over time, governments. People in some parts of the world are already reeling from the effects.

In Africa, with the fastest population growth of any continent on record, a combination of deforestation, overgrazing, soil erosion, and desertification contributed to a lowering of per capita grain production by some 17 percent from the historical peak in the late 1960s. The fall from an annual output of 155 kilograms per person in the late 1960s to 129 kilograms in the late 1980s has converted the continent into a grain-importing region, fueled the region's mounting external debt, and left millions of Africans hungry and physically weakened, drained of their vitality and productivity. In Latin America, rapid population growth and environmental degradation have contributed to a 7 percent fall in per capita grain production from the peak reached in 1981.

For the world, grain production per person climbed an impressive 40 percent between 1950 and 1984. Since 1984, it has fallen each year, dropping a record 13 percent during this 4-year span (Figure 12.1). Four-fifths of the 13 percent production decline was offset by reducing world carry-over stocks of grain from the equivalent of a record 101 days of world consumption in 1987 to under 60 days in 1989, not much more than "pipeline" supplies. The remaining one-fifth was absorbed by the 3 percent decline in world grain consumption per person that resulted from rising prices.

In describing this recent 4-year period, I do not mean to imply that the combination of population growth and environmental deterioration is solely responsible. Depressed world prices for farm commodities, ill-conceived farm policies, adverse weather, and even climate change may have contributed. Nor do I want to imply that this 4-year stretch is a new trend, but it does mean that it is becoming more difficult to systematically raise food production per person than it was prior to 1984.

Kilograms

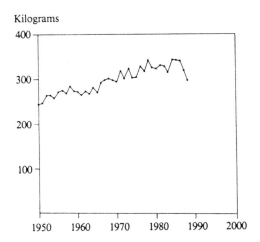

FIGURE 12.1 World grain production per capita, 1950-1988. SOURCE: U.S. Department of Agriculture.

THE AGRICULTURAL EFFECTS OF DEFORESTATION

For many years, reports on deforestation have described an 11-million-hectare annual loss of tropical forests, which was based on a 1980 Food and Agriculture Organization (FAO) survey. But when the Brazilian government, using information provided by satellites, reported that 8 million hectares of its Amazon rain forest were burned off in 1987, mostly to clear land for ranching and farming, it became clear that the Amazon is being destroyed far faster than previously thought.

The causes of deforestation vary, but land clearance for agriculture, as in Brazil, is the leading source of deforestation worldwide. In densely populated countries, such as India and Ethiopia, firewood gathering is more often responsible. An FAO study estimated that in 1980 some 1.2 billion of the world's people were meeting their firewood needs only by cutting wood faster than nature could replace it. In 1982, India's remaining forestland could sustain an annual harvest of only 39 million tons of wood, far below the estimated fuel wood demand of 133 million tons. The gap of 94 million tons was closed either by overcutting, thus compromising future firewood production, or by burning cow dung and crop residues, compromising future food production.

In Africa, the degree of imbalance between firewood demand and the unsustainable harvests of wood varies widely. For example, in both semiarid Mauritania and mountainous Rwanda, firewood demand is 10 times the sustainable yield of the remaining forests. In Kenya, the ratio is 5 to 1. In Ethiopia, Tanzania, and Nigeria, demand is 2.5 times the sustainable yield. And in the Sudan, it is roughly double.

Regardless of the reason for the tree cover losses, the consequences are usually the same. Soil organic matter declines, reducing the moisture storage capacity of the soil. Rainfall runoff increases. Percolation and aquifer recharge decrease. Soil erosion accelerates.

TABLE 12.1 Area Subject to Flooding in India as Deforestation Progresses

Year	Million Hectares
1960	19
1970	23
1980	40
1984	50

SOURCE: National Flood Commission and Center for Science and Environment, New Delhi, India.

Deforestation directly alters local hydrological cycles by increasing runoff and, perhaps less obviously, by affecting the recycling of rainfall inland. The former is now strikingly evident in the Indian subcontinent, where deforestation of the Himalayan watersheds is increasing rainfall runoff, leading to increasingly severe flooding. The data in Table 12.1 indicate that the area subject to annual flooding in India has expanded dramatically, more than doubling since 1960.

Accelerated runoff as a result of deforestation was evident in early September 1988, when two-thirds of Bangladesh was under water for several days. The 1988 flood, the worst on record, left 25 million of the country's 110 million people homeless, adding to the growing ranks of "environmental refugees."

Eneas Salati and Peter Vose have analyzed the effect of deforestation on the recycling of rainfall inland in the central Amazon. They point out that in a healthy stand of rain forest, about three-fourths of the rainfall is evaporated either directly from the soil and from the surface of leaves or from transpiration by plants, and roughly one-fourth runs off into streams, returning to the ocean (Table 12.2). Such high levels of cloud recharge have led ecologists to refer to tropical rain forests as "rain machines."

After deforestation, this ratio is roughly reversed, suggesting that as deforestation of the Amazon continues, the vigorous recycling of water inland from the Atlantic will weaken, leading to lower rainfall and a drying out of the western Amazon. Moisture left in the air when the westward-moving air masses are directed southward by the Andes into southern Brazil and the Chaco/Paraguay river regions becomes part of the rainfall cycle in major farming areas. If this is reduced, Salati and Vose believe it "might affect climatic patterns in agriculture in south central Brazil." In effect, efforts to expand beef production in the central Amazon could indirectly reduce rainfall and food production in the country's agricultural heartland to the south.

TABLE 12.2 Water Balance in Amazonian Watershed Near Manaus, Brazil

Path of Rainfall	Proportion of Rainfall (percent)
Evaporation of rainfall intercepted by vegetation and from forested soil	26
Transpiration from vegetation	48
Total evapotranspiration	74
Stream runoff	26
Total rainfall	100

SOURCE: Eneas Salati and Peter B. Vose, Amazon Basin: A System in Equilibrium, Science 225:138-144, 1984. Copyright (c) 1984 by the American Association for the Advancement of Science.

Although it has been the subject of little research, a similar situation exists in western Africa, where the interior region depends on rainfall that is recycled inland via the coastal rain forests. Although there is little or no research on deforestation and the recycling of rainfall into the continental interior, it is hard to see how the flow of moisture inland would not have been reduced by the extensive deforestation of the coastal tier of countries, stretching from Senegal through Nigeria. We do know that the isohyets have shifted steadily southward in the Sahelian region over the last 3 decades. Thousands of villages all across the southern edge of the Sahel have been abandoned in recent years. The number of Mauritanians living in Senegal and Mali may now exceed those remaining in Mauritania. Even the survival of some ancient cities, such as Timbuktu, is in question. The drying out and desertification of the Sahelian region probably account for the largest source of environmental refugees in the world today.

In addition to adversely affecting the hydrological cycle, deforestation can disrupt nutrient cycles as well, reducing the land's biological productivity. Drawing on field data from Ethiopia, World Bank ecologist Kenneth Newcombe reports that when land is without trees, mineral nutrients are no longer recycled from deep soil layers. As this nutrient cycle is breached, soil fertility begins to decline. As trees disappear, villagers begin to burn crop residues and animal dung for fuel. This in turn interrupts two more nutrient cycles: removing crop residues and diverting dung from fields both degrade soil structure and leave the land more vulnerable to erosion.

TABLE 12.3   Household Fuel Consumption in the Indian State of Madhya
Pradesh

Fuel	Quantity (million tons)
Cow dung	9.64
Firewood	9.47
Crop residues	6.93

SOURCE: Centre for Science and Environment, <u>The State of India's Environment, 1984-85</u>, New Delhi, 1985.

        Eventually cow dung and crop residues become the main fuel source.
Data gathered on household fuel use in the central Indian state of Madhya
Pradesh show this energy transition is well under way.   Cow dung has
edged out firewood as the principal household fuel, with the use of crop
residues not far behind (Table 12.3).   As this flow of nutrients from the
land into villages and towns continues, it drains the soil of its fertility, leaving farmers vulnerable to crop failure during even routine
dry seasons.
        If this process continues over an extended period, with no nutrient
replenishment, land productivity will decline to the point where families
can no longer produce enough food for themselves or their livestock, let
alone for markets.   A mass exodus from rural areas begins, often triggered by drought that could formerly have been tolerated.
        As deforestation directly and indirectly reduces soil organic matter
and moisture storage, it can lead to a new kind of drought--one that
results not from reduced rainfall but from the reduced ability of the
soil to store moisture.   This was among the concerns that led to the
convening in India of a national seminar in May 1986 entitled "Control of
Drought, Desertification and Famine."   Attended by nearly 100 profes-
sionals, the conference was concerned that the "temporary phenomenon of
meteorological drought in India has tended to be converted into the
permanent and pervasive phenomenon of desertification, undermining
biological productivity of soil over large parts of the country."   In a
radio address to the nation, Prime Minister Rajiv Gandhi recognized the
link with deforestation: "Continuing deforestation has brought us face-
to-face with a major ecological and economic crisis.   The trend must be
halted."
        Many of the costs of deforestation do not show up in national eco-
nomic accounts.   As nearby forests dwindle and disappear, women and
children travel further and work harder to meet minimal firewood needs.
Eventually, as in some villages in the Andes and the Sahel, firewood
scarcity reduces people to one hot meal per day.   Deforestation can not
only adversely affect food production, but it can also deprive people of
the fuel to cook what food they do produce.

109

OVERGRAZING AND THE LOSS OF GRASSLAND

Although the data for grassland degradation are even sketchier than are those for forest clearing, the trends are no less real. A United Nations study charting the mounting pressures on grasslands in nine countries in southern Africa shows that the capacity to sustain livestock populations is diminishing. This problem is noticeable throughout Africa, where livestock numbers have expanded nearly as quickly as the human population. In 1950, Africa had 219 million people and 295 million livestock. By 1987, the continent's human population had increased to 601 million and its livestock numbers to 539 million.

Because little grain is available for feeding them, the continent's 182 million cattle, 195 million sheep, and 162 million goats are supported almost entirely by grazing and browsing. Everywhere outside the tsetse fly belt, livestock are vital to the economy. But in many countries, herds and flocks are destroying the grassland resource that sustains them. The U.N. report on the nine countries in southern Africa observes that "for some countries, and major areas of others, present herds exceed the carrying capacity from 50 to 100 percent. This has led to a deterioration of the soil--thereby lowering the carrying capacity even more--and to severe soil erosion in an accelerating cycle of degradation."

Overgrazing gradually changes the character of rangeland vegetation and its capacity to support livestock. As degradation of rangeland continues, its capacity to carry cattle diminishes, leaving it to goats and sheep, which can browse the remaining woody plants. This shift in the composition of Africa's livestock herd has been particularly evident since 1970 (Table 12.4).

As grazing and wood-gathering increase in semiarid regions, the rapidly reproducing annual grasses replace perennial grasses and woody perennial shrubs. The loss of trees, such as the acacias in the Sahel, means less forage in the dry season, a time when the protein-rich acacia pods formerly fed livestock. Annual grasses that dominate the landscape are far more sensitive to stress than are perennials and may not germinate at all in dry years.

Fodder needs of livestock populations in nearly all Third World countries now exceed the sustainable yield of grasslands and other forage resources. In India, the demand for livestock fodder by the year 2000 is expected to reach 700 million tons, while the supply will total just 540 million tons. The National Land Use and Wastelands Development Council reports that in states with the most serious land degradation, such as Rajasthan and Karnataka, fodder supplies satisfy only 50 to 80 percent of needs, leaving large numbers of emaciated cattle. When drought occurs, hundreds of thousands of cattle die. In recent years, local governments in India have established fodder relief camps for cattle threatened with starvation, much as food relief camps are established for starving human populations.

Overgrazing is by definition a short-term phenomenon. Deteriorating grasslands that cannot sustain livestock populations cannot sustain the human populations that depend on them. Countless thousands of those who made a living from grazing their flocks and herds as recently as a decade

TABLE 12.4 Changes in Africa's Cattle, Sheep, and Goat Populations, 1950-1970 and 1970-1987

	Average Annual Change (percent)	
	1950-1970	1970-1987
Cattle	+2.15	+0.73
Sheep	+1.67	+1.55
Goats	+1.67	+1.85

SOURCE: U.N. Food and Agriculture Organization, <u>Production Yearbook</u>, Rome, various years.

or two ago now populate food relief camps in Africa or the squatter settlements that surround almost every major city in Africa and in the northern reaches of the Indian subcontinent.

THE COST OF SOIL EROSION

Soil erosion is a natural process, one that began as the first soil was formed when the earth was still young. Because new soil is being continuously formed from parent materials, erosion becomes an economic threat only when it exceeds the new rate of soil formation, which is typically estimated at 2 to 5 tons per acre per year.

As the demand for food has risen in recent decades, so have the pressures on the earth's soils. In the face of this continuing world demand for grain and the associated relentless increase in pressures on land, soil erosion is accelerating as the world's farmers are pressed into plowing highly erodible land and as traditional rotation systems that maintain a stable soil base are beginning to break down.

Throughout the Third World, increasing population pressure and the accelerating loss of topsoil seem to go hand in hand. Soil scientists S. A. El-Swaify and E. W. Dangler have observed that it is in precisely those regions with high population density that "farming of marginal hilly lands is a hazardous necessity. Ironically, it is also in those very regions where the greatest need exists to protect the rapidly diminishing or degrading resources." It is this vicious cycle, set in motion by the growing demands for human food, feed, fiber, and firewood, that makes mounting an effective response particularly difficult.

In other parts of the world, traditional cropping rotations that included nitrogen-fixing legumes permitted farmers to cultivate rolling land without losing excessive amounts of topsoil. Typical of these regions is the midwestern United States, where farmers traditionally used long-term rotations of hay, wheat, and corn. By alternating row crops, which are most susceptible to erosion, with cover crops, like hay, farmers kept soil erosion below the natural rate of new soil formation. As world demand for food soared after World War II, however, and as the cost of nitrogen fertilizer fell, farmers abandoned these rotations in favor of continuous row cropping.

An estimated one-third of the world's cropland is now losing topsoil at a rate that is undermining its long-term productivity. At the World-watch Institute, we estimate the worldwide loss of topsoil from crop-land, in excess of new soil formation, at 24 billion tons per year, roughly the amount of topsoil on Australia's wheatland.

When most of the topsoil is lost on land where the underlying forma-tion consists of rock or where the productivity of the subsoil is too low to make cultivation economical, it is abandoned. More commonly, however, land continues to be plowed even though most of the topsoil has been lost and even though the plow layer contains a mixture of topsoil and subsoil, with the latter dominating. Other things being equal, the real cost of food production on such land is far higher than on land where the topsoil layer remains intact.

Leon Lyles, an agricultural engineer with the U.S. Department of Agriculture (USDA), has provided perhaps the most comprehensive collec-tion of research results on the effect of soil erosion on land produc-tivity. Drawing on the work of U.S. soil scientists, both within and outside government, Lyles compared 14 independent studies, mostly under-taken in the Corn Belt states, to summarize the effects of a loss of 1 inch of topsoil on corn yields. His survey found that such a loss caused a reduction in yields ranging from 3.0 to 6.1 bushels per acre (Table 12.5). These 14 studies showed that the loss of 1 inch of topsoil reduced corn yields on 18 sites by an average of 6 percent.

Results for wheat, drawing on 12 studies, showed a similar relation-ship between soil erosion and land productivity. The loss of 1 inch of topsoil reduced wheat yields by 0.5 to 2.0 bushels per acre. In per-centage terms, the loss of 1 inch of topsoil reduced wheat yields an average of 6 percent, exactly the same as for corn.

Although there are few reliable data on the effect of soil erosion on land productivity for most countries, some insights into the relationship can be derived from these U.S. studies. Given the consistency of the decline in productivity across a wide range of soil types and crops, it would not be unreasonable to assume that a similar relationship between soil erosion and land productivity exists in other countries. Research on West African soils shows that a loss of 3.9 inches of topsoil, roughly half of an undisturbed topsoil layer, cuts corn yields by 52 percent. Yields of cowpeas, a leguminous crop, are reduced by 38 percent.

Because of the short-sighted way that one-third to one-half of the world's cropland is being managed, the soils on these lands have been converted from a renewable to a nonrenewable resource. Although the loss of topsoil does not show up in the national economic accounts or resource inventories of most countries, it is nonetheless a serious loss. Each year the world's farmers are trying to feed 86 million more people, but with 24 billion fewer tons of topsoil than the year before.

Grave though the loss of topsoil may be, it is a quiet crisis, one that is not widely perceived. Unlike earthquakes, volcanic eruptions, and other natural disasters, this human-made disaster is unfolding

TABLE 12.5 Effect of Topsoil Loss on Corn Yields

Location	Yield Reduction per Inch of Topsoil Lost Bushels per Acre	Percent	Soil Description
East Central, Illinois	3.7	6.5	Swygert silt loam
Fowler, Indiana	4.0	4.3	Fowler, Brookston, and Parr silt loams
Clarinda, Iowa	4.0	5.1	Marshall silt loam
Greenfield, Iowa	3.1	6.3	Shelby silt loam
Shenandoah, Iowa	6.1	5.1	Marshall silt loam
Bethany, Missouri	4.0	6.0	Shelby and Grundy silt loams
Columbus, Ohio	3.0	6.0	Celina silt loam
Wooster, Ohio	4.8	8.0	Canfield silt loam

SOURCES: Various reports cited in Leon Lyles, Possible Effects of Wind Erosion on Soil Productivity, Journal of Soil and Water Conservation, November/December 1975.

gradually. And it is unrecognized because the intensification of cropping patterns and the plowing of marginal lands that lead to excessive erosion over the long run can lead to production gains in the short run, thus creating the illusion of progress and a false sense of food security.

Although soil erosion is a physical process, it has numerous economic consequences affecting land productivity, economic growth, income distribution, food sufficiency, and long-term external debt. Ultimately it affects people. When soils are depleted and soils are poorly nourished, people are often undernourished as well. What is at stake is not merely the degradation of soil, but the degradation of life itself.

THE CHANGING POPULATION/LAND RELATIONSHIP

One of the most serious consequences of continuing population growth is the worldwide shrinkage in cropland per person. Between 1950 and 1981, the world grain area increased some 24 percent, reaching an all-time high (Figure 12.2). Since then it has fallen some 6 percent. That the world's cropland area would expand when the world demand for food was growing rapidly is not surprising. What is surprising--and worrying--is the recent decline. This is due partly to the systematic retirement of highly erodible land under conservation programs in the United States; partly to the abandonment of eroded land, as in the Soviet Union; partly

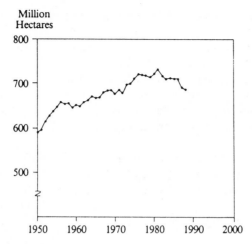

FIGURE 12.2 World harvested area of grain, 1950-1988. SOURCE: U.S. Department of Agriculture.

to the growing conversion of land to nonfarm uses, a trend most evident in densely populated Asia; and partly to cropland set aside to control production, as in the United States.

After the second surge in world grain prices between 1972 and 1973, farmers throughout the world responded to record prices by plowing more land. In the United States, they not only returned idled cropland to use, but they also plowed millions of acres of highly erodible land. Between 1972 and 1976, the U.S. area in grain climbed some 24 percent, but soil erosion apparently increased even more rapidly. By 1977, American farmers were losing an estimated 6 tons of topsoil for every 1 ton of grain they produced.

The United States is now in the fourth year of a 5-year program to convert at least 40 million acres of highly erodible cropland to either grassland or woodland before it loses most of its topsoil and becomes wasteland. As of today, some 28 million acres have been converted under 10-year contracts.

The Soviet Union, lacking such a program, has abandoned roughly 1 million hectares of grain land each year since 1977, leading to a 13 percent shrinkage in area. Abandonment on this scale suggests that inherent fertility may be falling on a far larger area, helping explain why the Soviets now lead the world in fertilizer consumption, using twice as much to produce a ton of grain as does the United States.

The conversion of cropland to nonfarm uses is also shrinking the cropland area. In China, one result of the past decade's welcome prosperity is that literally millions of villages are either expanding their existing dwellings or building new ones. And an industrial sector expanding at a rate of more than 12 percent annually since 1980 means the construction of thousands of new factories each year. Since most of China's 1.1 billion people are concentrated in its rich farming regions, new homes and factories are often built on cropland. This loss, combined with the shifts to more profitable crops, has reduced China's grain-growing area by 9 percent since 1976.

One country that can increase its cropland area somewhat in the short run is the United States. As recently as 1987, it was idling 50 million acres of cropland to control production. About half of this is being returned to production in 1989. The remainder could be returned to use in 1990. This will be substantially offset by the 40 million acres of highly erodible cropland, mentioned earlier, that is being withdrawn from production.

There are a few countries that are still steadily expanding their cultivated area. Brazil, for example, has nearly tripled its cultivated area since 1950, with most of the growth coming in the south and southeast outside the Amazonian basin. Although the expansion has slowed during the 1980s, further growth is a prospect over the remaining part of this century and beyond.

This modest short-term gain in the United States and the longer-term prospective gain in Brazil and elsewhere will expand the cropland base. It is unlikely, however, that these gains will offset the losses under way elsewhere. The prospect for the rest of this century is for no meaningful net addition to the world's cropland base.

Between 1950 and 1989, the world grain area per person declined from 0.23 hectares to 0.14 hectares, a shrinkage of 39 percent (Table 12.6). Assuming that the projected growth in population materializes with no net gain in world cropland over the next 2 decades, grain area per person will fall to 0.10 hectares per person by 2010, a further drop of 29 percent.

WATER FOR BREAD

Given the scarcity of new cropland, after mid-century many countries worked to raise land productivity by expanding the irrigated area. Between 1950 and 1980, the world irrigated area expanded from 94 million hectares to 249 million hectares, a 2.6-fold gain that closely paralleled the growth in food output (Table 12.7). After 1980, however, growth slowed dramatically.

Unfortunately, not all the irrigation expansion during the preceding 3 decades was sustainable. In recent years, the world's two leading food producers, the United States and China, have experienced unplanned declines in irrigated area. The U.S. irrigated area, which peaked in 1978, has shrunk some 7 percent since then, reversing several decades of growth. In addition to falling water tables, depressed commodity prices and rising pumping costs have contributed to the shrinkage.

Further declines are a prospect. In 1986, the USDA reported that more than one-fourth of the 21 million hectares of irrigated cropland was being watered by pulling down water tables, with the drop ranging from 6 inches to 4 feet per year. The water tables were falling either because the pumping exceeded aquifer recharge or because the water was from the largely nonrenewable Ogallala aquifer. Although water mining is an option in the short run, in the long run withdrawals cannot exceed aquifer recharge.

TABLE 12.6 World Grain Land, Total and Per Capita, 1950-1980, with Projections to 2010

Year	Total Grain Land (million hectares)	Per Capita Grain Land (hectares)	Change by Decade (percent)
1950	593	0.23	
1960	651	0.21	- 8
1970	673	0.18	-14
1980	724	0.16	-11
1990 (proj.)	720	0.14	-12
2000 (proj.)	720	0.12	-14
2010 (proj.)	720	0.10	-17

TABLE 12.7 World Irrigated Area, Total and Per Capita, 1950-1980, with Estimates for 1990

Year	Total Irrigated Cropland (million hectares)	Per Capita Irrigated Cropland (hectares)	Per Capita Change by Decade (percent)
1950	94	0.036	
1960	140	0.046	+28
1970	197	0.053	+15
1980	249	0.056	+ 6
1990 (proj.)	265	0.050	-11

SOURCE: Data for 1950 to 1980 adapted from W. R. Rangeley, Irrigation and Drainage in the World, paper presented at the International Conference on Food and Water, Texas A&M University, College Station, May 26-30, 1986; 1980 irrigated acreage prorated from 1970 and 1982 figures as cited in W. R. Rangeley, Irrigation--Current Trends and A Future Perspective, World Bank Seminar, February 1983; data for 1990 are author's projection.

In China, where the expansion peaked in 1978, irrigated area had shrunk by 2 percent by 1987. Under parts of the North China plain, in the region surrounding Beijing and Tianjin, the water table is dropping by 1 to 2 meters per year as industrial, residential, and agricultural users compete for dwindling supplies of water.

In the Soviet Union, the excessive use of water for irrigation takes the form of diminished river flows rather than falling water tables. Roughly one-third of the Soviet Union's irrigated cropland is centered around the Aral Sea in Central Asia. Irrigation diversions from the Syr-Darya and Amu-Darya, the two great rivers of the region that sustain the land-locked sea, have led to a 40 percent shrinkage in its area since 1960. Soviet scientists fear a major ecological catastrophe is unfolding as the sea slowly disappears. The dry bottom is now becoming desert, the site of sand storms that may drop on the surrounding fields up to one-half ton per hectare of a sand-salt mix, damaging the crops that water once destined for the sea is used to grow.

Competition between the countryside and cities for fresh water supplies is intensifying in many countries. Faced with absolute limits on the amount of fresh water available in the southern Great Plains and the southwestern United States, cities unable to afford new projects are buying irrigation water rights from farmers. In the competition between agricultural, residential, and industrial water users, it is agriculture that invariably surrenders water.

Irrigation systems are deteriorating in some countries. U.N. analysts estimate that close to 40 percent of the world's irrigated area is suffering from varying degrees of waterlogging and salinity. In many cases, this condition can be reversed, providing the capital is available for the installation of underground drainage systems. In other situations, however, the salt content of the water being used for irrigation is so high that there may not be any practical way of dealing with it, meaning that the irrigated land will eventually be abandoned.

There are still many opportunities for expanding the irrigated area, but given the losses that are occurring in some countries, the world is not likely to reestablish a trend of rapid, sustained growth in irrigated area like that from 1950 to 1980. In retrospect, this growth will probably have been unique. Any future gains in irrigated area may depend as much on gains in water use efficiency as on new supplies.

Between 1950 and 1980, world irrigated area expanded 2.6-fold, while population increased scarcely 1.7-fold, raising the amount of irrigation water used per person by 56 percent. This increase helped offset the effects of the shrinking cropland per person. But between 1980 and 1990, we estimate the irrigated area will increase by only 16 million hectares, far less rapidly than population, leading to a reduction in irrigated area per person of 11 percent. During the 1980s, for the first time since mid-century, the world is experiencing a shrinkage both in irrigation water and in cropland per person.

THE TECHNOLOGICAL PROSPECT

From the beginning of agriculture until around 1950, most of the growth in world food output came from expanding the cultivated area. As the frontiers disappeared around mid-century, farmers shifted to raising land productivity. Between 1950 and 1981, a period during which the cropland area expanded only modestly, roughly four-fifths of the growth in world food output came from raising productivity. During the 7 years

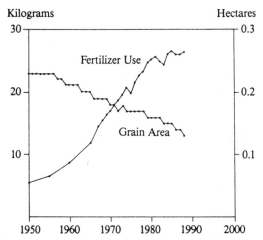

FIGURE 12.3 World fertilizer use and grain area per capita, 1950-1988.
SOURCE: U.S. Department of Agriculture.

since 1981, a period when the world cropland area declined, all growth
in output has come from land productivity gains. In effect, we now have
10,000 years of experience increasing food supplies primarily by expanding cultivated area and 4 decades by raising land productivity.

Between 1950 and 1984, the world's farmers raised their grain yield
per hectare from 1.1 tons to 2.3 tons, a remarkable feat. Four technol-
ogies--(1) chemical fertilizer, (2) irrigation, (3) high-yielding dwarf
wheats and rices, and (4) hybrid corn--accounted for most of the increase. Growth in fertilizer use has led the way. From 1950 through
1984, fertilizer use climbed from 14 million to 125 million tons, a gain
of more than 11 percent per year. Since then, growth in fertilizer use
has slowed dramatically as the growth in irrigated area has slowed, as
the yield response to fertilizer use has diminished, as commodity prices
have weakened, and as Third World debt has soared. In addition, many
financially pressed governments have reduced fertilizer subsidies. From
1984 to 1988, usage went from 125 million to 135 million tons, an annual
rise of only 2 percent.

Over the past generation, the world's farmers have successfully sub-
stituted fertilizer for land (Figure 12.3). In per capita terms, world
fertilizer use quintupled between 1950 and 1984, going from 5 to 26
kilograms and offsetting a one-third decline in grain area per person.
As varieties are improved, the response to fertilizer use continues to
rise, albeit slowly in recent years.

Eventually the rise of grain yield per hectare, like the growth of
any biological process in a finite environment, will conform to the
standard S-shaped growth curve. So, too, will the response to inputs,
such as fertilizer, that are responsible for the rise. The fertilizer
use curve in Figure 12.3 appears to be conforming to the S shape.

The ultimate constraints on the rise of crop yields will be imposed
by the upper limit of photosynthetic efficiency. Evidence that photosynthetic constraints may be emerging can be seen in the diminishing

returns on fertilizer use. Whereas 20 years ago the application of each additional ton of fertilizer in the U.S. Corn Belt added 15 or 20 tons to the grain harvest, today it may add only 5 to 10 tons. In analyzing recent agricultural trends in Indonesia, agricultural economists Duane Chapman and Randy Barker of Cornell University note that "while 1 kilogram of fertilizer nutrients probably led to a yield increase of 10 kilograms of unmilled rice in 1972, this ratio has fallen to about 1 to 5 at present."

If the response to additional fertilizer use is diminishing, what other technologies can continue to boost world food output in the way that the 10-fold increase in fertilizer use has since mid-century? Unfortunately no identifiable technologies are waiting in the wings that will lead to the quantum jumps in world food output produced by the four technologies outlined above.

There has been an overall loss of momentum in the growth in world food output. Although there are still many opportunities for expanding food output in all countries, it is becoming more difficult for some to maintain the rapid expansion in output that the growth of their population demands.

CLIMATE AND FOOD

Of all the global changes we have set in motion, climate change is potentially the most disruptive. Already suffering from slower growth in food output, the world is now confronted with the prospect of hotter summers. Farmers who have always had to deal with the vagaries of weather must now also contend with the uncertainty of worldwide climate change.

The drought- and heat-damaged U.S. grain harvest in 1988, which fell below consumption probably for the first time in history, dramatically illustrates how hotter summers may affect agriculture over the longer term in the United States and elsewhere. Grain production dropped to 196 million tons, well below the estimated 206 million tons of consumption (Table 12.8). This shortfall was filled by drawing down stocks. U.S. commitments to export close to 100 million tons during the 1988-1989 marketing year may be satisfied by exporting much of the remaining U.S. grain reserve. With a normal harvest, the United States typically harvests 300 million tons of grain, consuming 200 million tons and exporting roughly 100 million tons.

As noted earlier, climate change will not affect all countries in the same way. The projected rises by 2030 to 2050 of 1.5 to 4.5°C (3 to 8°F) are global averages, but temperatures are expected to increase much more in the middle and higher latitudes and more over land than over the oceans. They are projected to change little near the equator, whereas in the higher latitudes rises could easily be twice that projected for the earth as a whole. This uneven distribution will affect world agriculture disproportionately, since most food is produced on the land masses in the middle and higher latitudes of the northern hemisphere.

Given the constraints of time and space, discussion of how global warming will affect the food prospect will be limited to North America. Although they remain sketchy, meteorological models suggest that two of

TABLE 12.8 U.S. Grain Production, Consumption, and Exportable Surplus by
Crop Year, 1980-1988 (in million metric tons)

Year	Production	Consumption	Exportable Surplus from Current Crop
1980	268	171	+ 97
1981	328	179	+149
1982	331	194	+137
1983	206	182	+ 24
1984	313	197	+116
1985	345	201	+144
1986	314	217	+ 97
1987	277	215	+ 62
1988	196	206	- 10

SOURCES: U.S. Department of Agriculture, Economic Research Service,
World Grain Harvested Area, Production, and Yield 1950-1987, unpublished
printout, Washington, D.C., 1988; USDA, Foreign Agricultural Service,
World Grain Situation and Outlook, August 1988.

the world's major food-producing regions--the North American agricultural heartland and a large area of central Asia--are likely to experience a decline in soil moisture during the summer growing season as a result of higher temperatures and increased evaporation. If the warming progresses as the models indicate, some of the land in the U.S. western Great Plains that now produces wheat would revert to rangeland. The western Corn Belt would become semiarid, with wheat or other drought-tolerant grains that yield 40 bushels per acre replacing corn that yields over 100 bushels.

On the plus side, as temperatures increase the winter wheat belt will migrate northward, allowing winter strains that yield 40 bushels per acre to replace spring wheat yielding 30 bushels. A longer growing season would also permit a northward extension of spring wheat production in areas such as Canada's Alberta province, thus increasing that nation's cultivated area. On balance, though, higher temperatures and increased summer dryness will reduce the North American grain harvest, largely because of their negative impact on the all-important corn crop.

Drought, which afflicted most of the United States during the summer of 1988, is essentially defined as dryness. For farmers, drought conditions can result from lower than normal rainfall, higher than normal temperatures, or both. When higher temperatures accompany below-normal rainfall, as they did during the summer of 1988, crop yields can fall precipitously. Extreme heat can also interfere with the pollination of some crops. Corn pollination can easily be impaired by uncommonly high temperatures during the 10-day period in July when fertilization occurs.

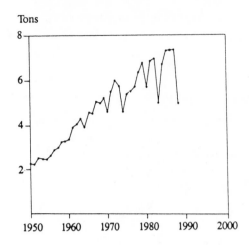

FIGURE 12.4 U.S. corn yield per acre, 1950-1988. SOURCE: U.S.
Department of Agriculture.

A rise in average temperatures will also increase the probability of
extreme short-term heat waves. If these occur at critical times--such
as the corn pollination period--they can reduce crop yields far more than
the relatively modest average temperature increase of a few degrees might
indicate.
    This vulnerability of corn, which accounts for two-thirds of the U.S.
grain harvest and one-eighth of the world's, can cause wide year-to-year
swings in the world grain crop. An examination of U.S. corn yields since
1950 shows five sharply reduced harvests over the last 38 years (Figure
12.4). The only pronounced drops before the 1980s came in 1970, from an
outbreak of corn blight, and in 1974, when a wet spring and late planting
combined with an early frost to destroy a part of the crop in the northern Corn Belt before it matured.
    Three harvests since 1950 have been sharply reduced by drought, all
in the 1980s. Each drop has been worse than the last. Compared with the
preceding year, the 1980 corn yield per acre was down by 17 percent, that
in 1983 was down by 28 percent, and that in 1988 by a staggering 34 percent.
    These three reduced harvests--in 1980, 1983, and 1988--each occurred
during 1 of the 5 warmest years of the last century. There may well be a
cause-and-effect relationship, but there is no way at this time to
conclusively link the drought-depressed U.S. harvests with a global
warming, since annual weather variability is so much greater than the
rise in average global temperatures measured during the 1980s. We do
know that the conditions experienced in the Corn Belt during the summer
of 1988 were similar to those described by the meteorological models as
the buildup of greenhouse gases continues. Although it is a scary
thought, if the drought and heat of 1988 are a sample of the hotter
summers to come, then the days of the North American breadbasket could be
numbered.

TWO SCENARIOS

What are the likely consequences of the recent slower growth in world food output and the global warming? Two widely asked questions define the two most common food scenarios. One is, What will the food situation be like if the world's weather this summer is "normal"? The other is, What if the United States experiences a severely drought-reduced harvest this summer, similar to that in 1988?

The answer to the first question is that even with normal weather, it may not be possible to rebuild depleted world grain stocks. This would mean that farmers are now having trouble keeping up with population growth and that for the foreseeable future the world will be living more or less hand-to-mouth, trying to make it from one harvest to the next.

The answer to the second question, which applies to future years as well if we cannot rebuild stocks, is that grain exports from North America would slow to a trickle. The world would face a food emergency. Never during the half century since America emerged as the world's breadbasket has it not had a large quantity of grain to export (Table 12.9). By September, there would be a frantic scramble for the comparatively meager exportable supplies of grain from France, Argentina, and Australia. There is no precedent by which to assess the impact of such a situation on grain prices. They could easily double, sending shock waves throughout the global economy that could destabilize national governments in low-income countries.

All available evidence indicates that the ranks of the hungry are expanding during the late 1980s, reversing the trend of recent decades. Uncertainties and stresses from a changing climate are now being overlaid upon an already tightening food situation. In the absence of a major commitment by governments to slow population growth and strengthen agriculture, food insecurity and the social instability associated with it will dominate the political landscape in many countries for years to come.

THE SOCIAL EFFECTS

As noted earlier, per capita grain production is now declining in two regions of the world. In Africa it has fallen 17 percent over the last 2 decades, and in Latin America it has fallen 7 percent from its all-time high in 1981. This sustained decline in grain output per person, which is likely to continue in these two regions, could spread to other regions during the 1990s.

In real terms, world grain prices were at an all-time low in early 1987, having fallen slowly, albeit irregularly, for many decades. But in a 1-year span between July 1987 and July 1988, world grain prices went up by roughly one-half, where they have since remained. As of March 1989, wheat prices were up from July 1987 by 62 percent, rice prices by 34 percent, and corn prices by 56 percent (Table 12.10).

Rising grain prices combined with falling incomes in many heavily indebted Third World countries pose a dilemma. Higher prices are needed to stimulate output and encourage additional investment by farmers. But

TABLE 12.9 The Changing Pattern of World Grain Trade, 1950-1988 (in million metric tons)

Region	1950	1960	1970	1980	1988[1]
North America	+23[2]	+39	+56	+131	+119
Latin America	+ 1	0	+ 4	- 10	- 11
Western Europe	-22	-25	-30	- 16	+ 22
E. Europe and Soviet Union	0	0	0	- 46	- 27
Africa	0	- 2	- 5	- 15	- 28
Asia	- 6	-17	-37	- 63	- 89
Australia and New Zealand	+ 3	+ 6	+12	+ 19	+ 14

[1]Preliminary.
[2]Plus sign indicates net exports; minus sign, net imports.

SOURCES: U.N. Food and Agriculture Organization, Production Yearbook, Rome, various years; U.S. Department of Agriculture, Foreign Agricultural Service, World Rice Reference Tables and World Wheat and Coarse Grains Reference Tables, unpublished printouts, Washington, D.C., June 1988.

on the demand side of the equation, the world's poor cannot cope with higher prices. Perhaps a billion or more of the world's people are already spending 70 percent of their income on food. If the price of grain rises dramatically, they will be unable to adjust. They will be forced to tighten their belts, but they do not have any notches left.

The social effect of higher grain prices is much greater in developing countries than in industrial ones. In the United States, for example, a $1 loaf of bread contains roughly 5 cents worth of wheat. If the price of wheat doubles, the price of the loaf will increase only to $1.05. In developing countries, however, where wheat is purchased in the market and ground into flour at home, a doubling of wholesale grain prices translates into a doubling of bread prices. For those who already spend most of their income on food, such a rise can drive consumption below the survival level.

Even before the recent grain price rises, the social effects of agricultural adversity were becoming highly visible throughout Africa. In mid-1988, the World Bank, using data through March 1986, reported that "both the proportion and the total number of Africans with deficient diets have climbed and will continue to rise unless special action is taken."

In Africa, the number of "food insecure" people, defined by the World Bank as those not having enough food for normal health and physical activity, now totals over 100 million. Some 14.7 million Ethiopians,

TABLE 12.10 World Grain Prices, March 1989 Compared with July 1987 (in U.S. dollars)

Grain	Price		Change (percent)
	July 1987	March 1989	
Wheat	2.85[a]	4.63[a]	+62
Rice	2.12[b]	2.84[b]	+34
Corn	1.94[a]	3.02[a]	+56

[a]Per bushel.
[b]Per ton.

SOURCE: International Monetary Fund, International Financial Statistics, Washington, D.C.

one-third of the country, are undernourished. Nigeria is close behind, with 13.7 million undernourished people. The countries with 40 percent or more of their populations suffering from chronic food insecurity are Chad, Mozambique, Somalia, Uganda, Zaire, and Zambia. The World Bank summarized the findings of its 1988 study by noting that "Africa's food situation is not only serious, it is deteriorating."

In its 1988 report The Global State of Hunger and Malnutrition, the U.N. World Food Council states that the number of malnourished preschoolers in Peru increased from 42 percent to 68 percent between 1980 and 1983. Infant deaths have risen in Brazil during the 1980s. If recent trends in population growth, land degradation, and growth in external debt continue, Latin America's decline in food production per person will almost certainly continue into the 1990s, increasing the number of malnourished people. The World Food Council summarized its worldwide findings by noting that "earlier progress in fighting hunger, malnutrition and poverty has come to a halt or is being reversed in many parts of the world."

When domestic food production is inadequate, the ability of countries to import becomes the key to food adequacy. During the late 1980s, low-income grain-deficit countries must contend not only with an increase in grain prices, but also in many cases with unmanageable external debt, which severely limits their expenditures on food imports. The World Bank nutrition survey of Africa was based on data through 1986; since then, conditions have deteriorated further as world grain prices have climbed.

Time and space constraints have limited this assessment to global changes that will affect food production in the near term. Others, such as rising sea level and stratospheric ozone depletion, will exert a greater influence over the longer term. Increasing ultraviolet radiation is of particular concern because it could adversely affect both the oceanic food chain and the yield of the more sensitive crops, such as soybeans, the world's leading protein crop.

A NEW FOOD ERA

Nearly all the global changes summarized in this paper are affecting the food prospect negatively. In the preceding pages, I have outlined the effects of many of these changes, including soil erosion, deforestation, increased rainfall runoff, decreased recycling of rainfall inland, waterlogging and salting of irrigation systems, falling water tables, grassland degradation, and shrinking cropland area and irrigation water supplies per person. In many countries, these negative influences on agriculture are now overriding the contribution of new investment and the adoption of more productive technologies designed to raise food per capita production.

The disturbing conclusion of this analysis is that the year 1984 may be a fault line separating two distinct eras in the world food economy. Between 1950 and 1984, the world was able to systematically raise grain production per person, lifting it some 40 percent, or more than 1 percent per year. In the new era, dating from 1984, we may not be able to count on systematic worldwide gains in per capita food output without a massive reordering of priorities. Indeed, it could take many years merely to regain the 13 percent loss in per capita grain production since 1984.

In the new era, the food prospect may depend as much on the ability of energy policymakers to trim carbon emissions as on the ability of agricultural policymakers to stimulate food output. For if energy policymakers do not act quickly, they could leave farmers with an impossible task of trying to feed 86 million more people per year in the midst of potentially convulsive climate change.

And in the new circumstances, where expansion of food output is more difficult, achieving an acceptable balance between food and people may depend more on family planners than on farmers. The issue is not whether population growth will eventually slow; it will. The only question is whether it will slow because we quickly move toward smaller families or because we let hunger and rising death rates check population growth, as they now are doing in some countries in Africa.

The gap between what we need to do to protect our environmental support systems and what we are doing is widening. Unless we redefine security, recognizing that the principal threats to our future come less from the relationship of nation to nation and more from the deteriorating relationship between ourselves and the natural systems and resources on which we depend, then the human prospect as we enter the twenty-first century could be a bleak one. If we do not act quickly, there is a risk that environmental deterioration and social disintegration could begin to feed on each other.

13

IMPACTS OF FUTURE SEA LEVEL RISE

James M. Broadus

Economists like to tell a story about the famous Gilded Age financier, J. P. Morgan. It seems a yuppie of those days found himself seated next to the old wizard and decided to play for some free advice. "What do you think the market will do?" he asked. Morgan looked at him sternly, glanced about, and leaned closer to whisper, "Fluctuate."

Exactly the same can be said of sea level, but with even greater certainty and a much longer record of experience. Changes in sea level are recorded on epochal scales as well as observed in real time (Figure 13.1). They are associated with regional tectonics, mesoscale oceanographic features, local land subsidence and erosion, and tides, waves, and ripples. We can speak of changes in average global eustatic sea level (determined by the volume of the ocean) and of changes in local relative sea level (local average height of the sea level relative to the land). In some places, local relative sea level has been rising for some time (Figure 13.2a). In other places it has been falling (Figure 13.2b). What changes have been taking place recently in average global eustatic sea level we do not really know. What changes to expect in coming decades and what their implications are for us now are the issues at hand.

The presumption in this case is that global warming, driven by the buildup of greenhouse gases from human activities, will increase the volume of water in the oceans (through a combination of thermal expansion and melting ice) and lead to sea level rises throughout the world (UNEP, 1986; Titus, 1986; Robin, 1987). Before discussing the status of current efforts to estimate the impacts of future sea level change, it is useful to consider a dozen brief observations that condition the exercise.

1. The expected physical impacts of rising sea level (Titus, 1986, 1987; Bird, 1986; Bruun, 1986; Park et al., 1986; EPA, 1988) include:

o inundation of low coastal lands;
o relocation or destruction of coastal wetlands;
o shoreline erosion and beach loss;
o exacerbated exposure to storm surge and flooding (Figure 13.3); and
o increased salinity of rivers, bays, and aquifers.

126

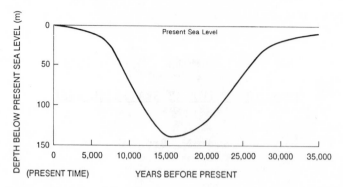

FIGURE 13.1 Recent changes in sea level. Small-scale fluctuations are
not shown. (Adapted from D. A. Ross. 1977. <u>Introduction to Oceanog-
raphy</u>, Prentice-Hall, Englewood Cliffs, N.J.)

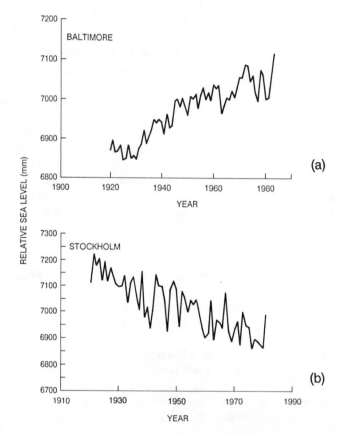

FIGURE 13.2 Tide-gauge records from the early twentieth century indicating (a) rising local relative sea level at Baltimore on North
American East Coast and (b) falling local relative sea level at Stockholm
in Scandinavia, as the land rises in "rebound" from weight of receding
glaciers. (Courtesy D. G. Aubrey and A. R. Solow, Woods Hole Oceanographic Institution.)

2.   Coasts are dynamic, ever-shifting places (Pilkey et al., 1982; Bird and Schwartz, 1985; NRC, 1987). Build too close and you will pay a price (Figure 13.4).

3.   Humans have long experience with coastal natural disasters (Figure 13.5). Sea level rise will only add to the problem, not create it.

4.   Impacts depend on the magnitude and pace of sea level rise, and this is _very_ uncertain (IAPSO, 1985).

5.   Sea level impacts depend on local relative sea level change, which varies from place to place. We still cannot distinguish local from global sea level change (Barnett, 1984; Solow, 1987), and in any case global trends are less pertinent for impact assessments and policy planning.

6.   Most efforts to determine a long-term trend in global average sea level change suggest a gradual rise of 1 to 2 mm/yr over the past century (Barnett, 1983). The tide-gauge data from which such estimates are derived are extremely patchy and variable in quality, and the data are extremely "noisy" at best. Also, the statistical techniques employed are unsettled and require further refinement (Solow, 1987; Gornitz and Solow, 1989). Evidence of an acceleration in the rate of sea level rise in response to an enhanced greenhouse warming (Robin, 1987) has not been detected, with the most likely change point identified so far coming in the late nineteenth century (Gornitz and Solow, 1989). That would be far too early to be ascribed to the modern buildup of greenhouse gases.

7.   The projections we do have for future sea level changes, on which estimates of future impacts must be based, are only scenarios describing hypothesized future conditions (Hoffman et al., 1986). These scenarios usually span a reported range, including low, medium, and high cases, for example; but typically they are not associated with estimates of relative likelihood.   What is needed are _probabilistic forecasts_, so that estimated future impacts can be weighted by their likelihood. In fact, there is currently no valid quantitative estimate of the extent and size of expected sea level increases.

8.   Impacts depend on human responses (Schelling, 1983; Bird, 1986; Bruun, 1986; Gibbs, 1986; NRC, 1987; Broadus, 1989). People are good at incremental adaptation, risk management, and technological advance.

9.   Economic impacts are sensitive to property values and _durable fixed capital_. In market economies, property values of land provide a workable approximation of the present value of the future flow of economic services supplied exclusively by the land. Property values can thus be used in estimating the value of projected land loss. As property values increase, so does the economic cost of potential inundation. Labor employed in many economic activities taking place in exposed areas can relocate, and so inundation will only impose a cost of adjustment and some penalty for moving to "next-best" productive employment. For physical capital in exposed areas (such as tools, cars, trailers, houses, wharves, shops, and factories), some can be moved and some will have worn out anyway before inundation. Therefore, it is really the value of long-lived or durable fixed capital (such as power plants, waste treatment facilities, nuclear waste disposal emplacements, highways, and port infrastructure) that must be reckoned into estimates of future losses.

128

FIGURE 13.3 Twelve-foot storm surge comes ashore at Galveston, Texas, during Hurricane Frederick in September 1979. (Courtesy of the National Oceanic and Atmospheric Administration.)

FIGURE 13.4 Cape Cod house tumbles into the sea in 1988 after a break in offshore barrier spit increased shoreline erosion at Chatham, Massachusetts. (Courtesy of Cape Cod Times.)

FIGURE 13.5 "Lamentable News out of Monmouthshire:" Old woodblock illustration of fourteenth-century coastal flooding in England. (Courtesy of University of East Anglia, Norwich, United Kingdom.)

The more the useful lifetime of such installations extends into the period of expected inundation, the greater the potential impact.

    10. The present economic value of future impacts depends critically on the social rate of discount. This is the factor by which future costs and benefits are reduced to make them comparable with present costs and benefits. It makes good economic sense to apply some discount rate to future values, because people do tend to value a current payment or benefit more heavily than a nominally equal payment at a future date. That is smart because to do otherwise would be to ignore the additional earnings that could be gained from the current payment (for instance, through investing it for compounded interest payments) in the time before the future payment becomes due. Selecting the appropriate discount rate to apply in practice, however, is quite difficult (Lind et al., 1982). It involves strong judgments about the preferences of society and encounters serious ethical complications when intergenerational effects are at stake. Higher discount rates depress the estimated cost of future sea level impacts. Lower discount rates make them appear larger.

    11. Impacts (and responses) will change with changes in tastes, preferences, and relative scarcity. Not so long ago, for example, wetlands were widely considered to be little more than useless wasteland, better to be filled and built on. With growing environmental awareness, increased scientific understanding of their functions, and changing aesthetic appreciation, wetlands are assuming a much loftier status. A similar effect might well have been expected anyway as wetlands became

TABLE 13.1 Nationwide Impacts of Sea Level Rise

Response Scenario	Hypothetical Sea Level Rise		
	50 cm	100 cm	200 cm
If densely developed areas are protected			
Shore protection costs (billions of dollars)	32-53	73-111	169-309
Dry land lost (mi^2)	2,200-6,100	4,100-9,200	6,400-13,500
Wetlands lost (%)	20-45	29-69	33-80
If no shores are protected			
Dry land lost (mi^2)	3,300-7,300	5,100-10,300	8,200-15,400
Wetlands lost (%)	17-43	26-66	29-76
If all shores are protected			
Wetlands lost (%)	38-61	50-82	66-90

SOURCE: EPA (1988).

scarcer under the pressure of various assaults. Changing tastes and public preferences are also altering favored responses. Where "hard" defensive measures such as dikes or seawalls might once have been selected, "soft" responses such as setback requirements and abandonment seem to be growing in popularity.

12. An interesting trade-off has been identified between response measures aimed at preservation of wetlands by allowing them to migrate with rising sea level and defensive measures aimed at protection of developed dry land behind them (EPA, 1988; Titus, 1988; Titus et al., 1984). EPA estimates suggest, for example, that up to 61 percent of the nation's wetlands might be lost to a 50-cm sea level rise if all coastal dry land were defended, while as little as 20 percent would be sacrificed if only densely developed areas were protected (Table 13.1).

Most attempts to assess the economic impact of future sea level rise employ a kind of "coloring book" approach. First, a scenario is selected that describes a hypothetical rise, often within a specified time period. The area subject to inundation from such a rise is then identified with topographic information and "colored in" on a map (Figure 13.6). The impact analyst then applies cost-benefit techniques, often based on property values, to estimate the potential loss in economic terms. The exercise is usually repeated for a range of scenarios in an attempt to bound the actual value.

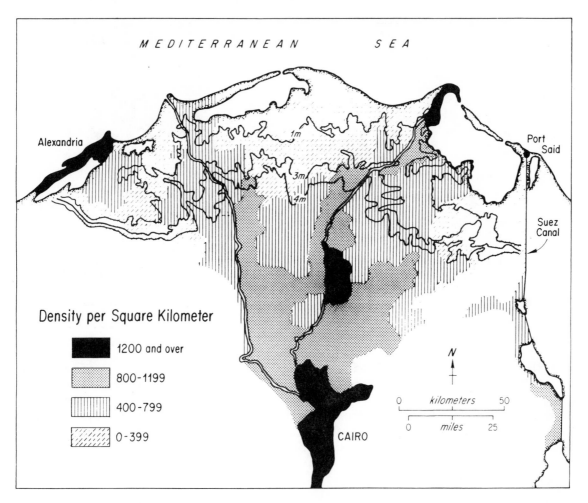

FIGURE 13.6 Inundation scenarios, superimposing 1-m, 3-m, and 4-m relative sea level rises on population density for the Nile Delta, Egypt
(Milliman et al., 1989).

The scenarios are usually selected from a range of projections that
have been developed on the basis of crude models of the relationship
between atmospheric temperature and sea level change (Figure 13.7).
Often the scenarios are adapted from these ranges to account for local
conditions such as land subsidence (Barth and Titus, 1984). Recent
estimates by Raper et al. (1988), which center on a "best-guess" range of
a 12- to 18-cm increase in eustatic sea level by 2030, are more moderate
than some earlier projections but still fall roughly within the low to
medium range proposed by Hoffman et al. (1986). A relatively rapid departure from the apparent background trend is seen for all the scenarios
in Figure 13.7. Recall that such an acceleration has yet to be detected
in the statistical record, and Raper et al. (1988) are careful to include
a zero increase within their broader range of possible changes.

A selection of economic impact estimates are reported in Table 13.2.
These focus on a variety of locales, time frames, and sea-level-rise

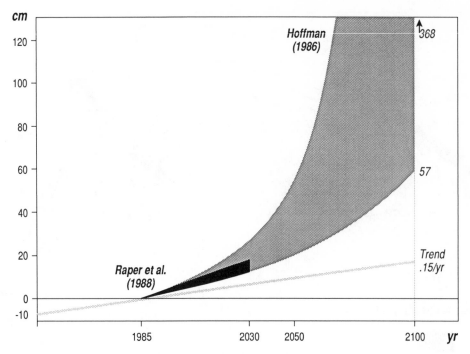

FIGURE 13.7 Typical range of sea-level-rise scenarios from Hoffman et
al. (1986) (in dark gray stippled region) compared to shorter-term
projections (black zone) by Raper et al. (1988) and contrasted with
apparent trend over past century (light grey line) of about 0.15 cm/yr.
Notice degree of acceleration above trend that is required if projections
and scenarios are to be realized.

scenarios. They also employ somewhat different estimation techniques
using different assumptions about economic growth, human responses, and
the social rate of discount. The original sources also reported their
cost estimates in differing constant dollar terms (corrected for in-
flation), but in Table 13.2 all are expressed in constant 1987 dollars
for easy comparison.

Perhaps the most careful and useful impact assessment to date is that
of Gibbs (1984) for a range of scenarios for Charleston, South Carolina,
and Galveston, Texas. Low and medium scenarios for the year 2075 are
shown in Table 13.2. These estimates show the great sensitivity to rate-
of-increase assumptions, jumping from a present value (using a 3 percent
discount rate) cost estimate of $760 million for the low case in Galve-
ston (92.4 cm) to $1.3 billion for the medium case (164.5 cm). (Note
too, however, that both the "low" and "medium" cases are in the upper
portion of the range of scenarios given by Hoffman et al. (1986).)

The Gibbs estimates in Table 13.2 also illustrate the remarkable
influence of the choice of discount rate. For example, moving from a
3 percent discount rate to a 10 percent discount rate reduces the higher
Charleston cost estimate of $2.6 billion to a mere $68 million.

133

TABLE 13.2 Selected Cost Estimates Suggesting Economic Impact of Sea Level Rise at Various Sites (millions of constant 1987 dollars)

Source	Social Rate of Discount (%)	Site	Date	Hypothetical Sea Level Rise (cm)	Cost
Gibbs (1984)	3	Charleston, S.C.	2075	87.6	1,712
				159.2	2,616
Gibbs (1984)	10	Charleston, S.C.	2075	87.6	27
				159.2	68
Gibbs (1984)	3	Galveston, Tex.	2075	92.4	760
				164.5	1,322
Gibbs (1984)	10	Galveston, Tex.	2075	92.4	14
				164.5	27
Schneider and Chen (1980)	0	United States	2130	760.0	555,555
Broadus (1989)	g[a]	Nile Delta	2050	100.0	521
Broadus (1989)	g[a]	Bangladesh	2050	100.0	417
Yohe (1988)	0	Long Beach Island, N.J.	2050	100.0	345
Yohe (1988)	0	Long Beach Island, N.J.	2100	200.0	1,942
Wilcoxen (1986)	3	San Francisco Sewer Transport	2100	177.4	74

[a]Social rate of discount equals economic growth rate, g.

Similarly for Galveston, the $760 million present-value lower cost estimate using a 3 percent discount rate falls to only $14 million if a 10 percent discount rate is used instead.

Schneider and Chen (1980) attempted an estimate for national losses to future sea level rise, using a scenario of a 760-cm (25-ft) rise by the year 2130. Based on 1971 property values and implicitly assuming a zero rate of discount, they calculated total property losses of over

$0.5 trillion ($200 billion in 1971 dollars) and speculated that ancillary damages might double the total. Their estimate is reported here only as an instructive historical curiosity. No student of the subject would now take the 25-ft projection seriously.

National impact estimates have also been attempted for Egypt and Bangladesh (Broadus et al., 1986; Broadus, 1989; Milliman et al., 1989). Again, areas subject to future inundation were delineated for various scenarios. Then demographic information, land use patterns, and national income accounts were used to estimate the current scale of economic activity taking place within the potentially affected areas. Thus, 7 percent of habitable land and 5 percent of current population were estimated to occupy the area subject to inundation by a 1-m rise in Bangladesh, with 12 percent of habitable land and 14 percent of population in the potentially affected area of Egypt (Broadus, 1989).

Although property value data were not available (and would be of dubious use in any case because of market distortions), a crude attempt was made to extend these estimates to present-value cost estimates. Using strong but reasonable parametric assumptions (discount rate equals economic growth rate, most of the productive land inundated is agricultural, agricultural rents are a modest proportion of total agricultural output, and absent mitigating responses), it was estimated that a "worst-case" relative sea level rise of 1 m by 2050 could impose a cumulative loss in present-value terms of roughly $0.5 billion in both Bangladesh and Egypt (Broadus, 1989).

In work currently under way, Yohe (1988) has examined the case of Long Beach Island, New Jersey. Using a sample of actual property values in strips extending across the island and applying a range of scenarios for rates of sea level rise, he has estimated the losses that might be incurred by "not holding back the sea." For example, a 1-m sea level rise by 2050 is estimated to threaten some $345 million in property values, while a 2-m rise by 2100 could wipe out some $2 billion in property values, essentially the entire current property value of the island. Yohe (1988) implicitly assumes a zero rate of discount.

In an interesting and somewhat different economic impact assessment, Wilcoxen (1986) estimated the additional cost that sea level rise could impose on the lifetime cost of a major sewer transport installation near San Francisco. Sea level rise had not been factored into the original engineering cost estimates for the project. Considering various sea level scenarios and weighting them by a subjective likelihood of realization, Wilcoxen projected $74 million as the present value (at a 3 percent rate of discount) of cumulative expenditures on beach nourishment that might be required to protect the system from erosion and wave attack over its 100-year planned life.

Wilcoxen's (1986) effort to form an aggregate scenario by weighting each of his various other scenarios with a "rough probability of occurrence" is noteworthy (although how the rough probabilities were arrived at was not reported). It results in a true estimate of expected cost, rather than the usual "certainty equivalent" assumption implied in the estimates derived from all the other scenarios reported in Table 13.2. It also calls attention again to the need for probabilistic forecasts of future sea level changes rather than the more questionable scenarios

FIGURE 13.8
"Going to School."
Boston flood.
Harper's Weekly,
February 27, 1886.

currently in use. This is an important area for further progress in our
ability to understand the present implications of future sea level
change.

Progress is being made. Major survey efforts to extend and improve
estimates of potential impact have been mounted by the U.S. Environmental
Protection Agency and by the United Nations Environment Programme, among
others. Increasingly, impact assessments are being linked to consideration of prospective response strategies. In the physical sciences, rapid
progress is being made in statistical analysis, modeling techniques, and
observational capability. Headway has been made in the effort to adjust
trends in long-term sea level change for background effects from global
tectonism (Peltier and Tushingham, 1989; Emery and Aubrey, 1985), and
satellite-based radio altimetry will permit direct and precise observation of changes in eustatic sea level. Fundamental knowledge of physical coastal and oceanographic processes is growing, and recognition of
the presence and importance of these processes to human activities is
becoming more widespread.

The immensity of our uncertainty about future sea level change should
not be understated, and a sustained commitment of effort and resources is
required to maintain our progress in reducing that uncertainty. Like the
children in a flooded Boston of a century past, we are still going to
school (Figure 13.8). But we are fast learners. On balance, it appears,

we are getting better at understanding and addressing the problem of sea
level rise faster than it is getting worse.

ACKNOWLEDGMENTS

The author is indebted for instruction and guidance on this subject
to John Milliman, Andy Solow, Dave Aubrey, K. O. Emery, Eric Bird, Jim
Titus, and Gary Yohe, although all errors and misjudgments are his own.
Financial support from The Pew Charitable Trusts and from the U.S.
Environmental Protection Agency is gratefully acknowledged. W.H.O.I.
Contribution No. 7181.

REFERENCES

Barnett, T.P. 1984. The estimation of "global" sea level change: A
    problem of uniqueness. J. Geophys. Res. 89(C5):7980-7988.
Barnett, T.P. 1983. Recent changes in sea level and their possible
    causes. Climate Change 5:15-38.
Barth, M.C., and J.G. Titus (eds.). 1984. Greenhouse Effect and Sea
    Level Rise. New York: Van Nostrand Reinhold Company.
Bird, E.C.F. 1986. Potential effects of sea level rise on the coasts of
    Australia, Africa and Asia. In Titus, J.G., (ed.), 1986.
Bird, E.C.F., and M.L. Schwartz (eds.). 1985. The World's Coastline.
    Stroudsburg: Van Nostrand Reinhold.
Broadus, J.M. 1989. Possible impacts of and adjustments to sea level
    rise: The cases of Bangladesh and Egypt. In T. Wigley and R.
    Warrick (eds.), The Effects of Climate Change on Sea Level, Severe
    Tropical Storms and their Associated Impacts, Climatic Research Unit,
    University of East Anglia, Norwich, U.K., in press.
Broadus, J.M., J.D. Milliman, S.F. Edwards, D.G. Aubrey, and F. Gable.
    1986. Rising sea level and damming of rivers: Possible effects in
    Egypt' and Bangladesh. In Titus, J.G. (ed.), 1986.
Bruun, P. 1986. Worldwide impact of sea level rise on shorelines. In
    Titus, J.G. (ed.), 1986.
Emery, K.O., and D.G. Aubrey. 1985. Glacial rebound and relative sea
    levels in Europe from tide-gauge records. Tectonophysics 120:239-
    255.
Environmental Protection Agency (EPA). 1988. Sea level rise. In Draft
    Report to Congress on the Potential Effects of Global Climate Change
    on the United States.
Gibbs, M.J. 1986. Planning for sea level rise under uncertainty: A
    case study of Charleston, South Carolina. In Titus, J.G. (ed.),
    1986.
Gibbs, M.J. 1984. Economic analysis of sea level rise: Methods and
    results. In Barth, M.C., and J.G. Titus (eds.), 1984.
Gornitz, V., and A. Solow. 1989. Observations of long-term tide-gauge
    records for indications of accelerated sea-level rise. Proceedings
    of the Department of Energy Workshop on Greenhouse Gas Induced

Climatic Change, Amherst, Mass., May 1989. Washington, D.C.:
    Department of Energy, forthcoming.

Hoffman, J.S., J. Wells, and J.G. Titus. 1986. Future global warming
    and sea level rise. In C. Sigbjarnarson (ed.), Iceland Coastal and
    River Symposium. Reykjavik, Iceland: National Energy Authority.

IAPSO Advisory Committee on Tides and Mean Sea Level. 1985. Changes in
    relative mean sea level. EOS, Transaction, Am. Geophys. U.
    66(45):754-756.

Lind, Robert C., K.J. Arrow, G.R. Corey, P. Dasgupta, A.K. Sen, T.
    Stauffer, J.E. Stiglitz, J.A. Stockfisch, and R. Wilson. 1982.
    Discounting for Time and Risk in Energy Policy. Washington, D.C.:
    Resources for the Future, Inc.

Milliman, J.D., J.M. Broadus, and F. Gable. 1989. Environmental and
    economic impact of rising sea level and subsiding deltas: the Nile
    and Bengal examples. Ambio, in press.

National Academy of Sciences (NAS). 1983. Changing Climate.
    Washington, D.C.: National Academy Press.

National Research Council (NRC). 1987. Responding to Changes in Sea
    Level: Engineering Implications. Washington, D.C.: National
    Academy Press.

Park, R.A., T.V. Armentano, and C.L. Cloonan. 1986. Predicting the
    effects of sea level rise on coastal wetlands. In Titus, J.G. (ed.),
    1986.

Peltier, W.R., and A.M. Tushingham. 1989. Global sea level rise and the
    greenhouse effect: Might they be connected? Science 244:806-810.

Pilkey, O., et al. 1982. Saving the American Beach: A Position Paper
    by Concerned Coastal Geologists. Savannah: Skidaway Institute of
    Oceanography.

Raper, S.C.B., R.A. Warrick, and T.M.L. Wigley. 1988. Global sea level
    rise: Past and future. In J.D. Milliman and S. Sabhasri (eds.),
    Proceedings, SCOPE Working Group on Rising Sea Level and Subsiding
    Coastal Areas, 9-13 November 1988, Bangkok, Thailand. New York:
    John Wiley & Sons, in press.

Robin. G. 1987. Projecting the rise in sea level caused by warming of
    the atmosphere. In Bolin, B., B.R. Döös, J. Jäger, and R.A. Warrick
    (eds.), The Greenhouse Effect, Climatic Change, and Ecosystems. New
    York: John Wiley & Sons.

Schelling, T.C. 1983. Climatic change: Implications for welfare and
    policy. Pp. 449-482 in NAS, 1983.

Schneider, S.H., and R.S. Chen. 1980. Carbon dioxide warming and
    coastline flooding: Physical factors and climatic impact. Ann. Rev.
    Energy 5:107-140.

Solow, A.R. 1987. The application of eigenanalysis to tide-gauge
    records of relative sea level. Continental Shelf Research 7(6):629-
    641.

Titus, J.G. (ed.). 1988. Greenhouse Effect, Sea Level Rise, and Coastal
    Wetlands. Washington, D.C.: Environmental Protection Agency.

Titus, J.G. 1987. Causes and effects of sea level rise. In Preparing
    for Climate Change, Proceedings of the First North American
    Conference on Preparing for Climate Change: A Cooperative Approach,

October 27-29, 1987, Washington, D.C. Rockville, Md.: Government
Institutes, Inc.

Titus, J.G. (ed.). 1986. Effects of Changes in Stratospheric Ozone and
Global Climate, Volume 4: Sea Level Rise. U.S. Environmental
Protection Agency, Washington, D.C.

Titus, J.G., T.R. Henderson, and J.M. Teal. 1984. Sea level rise and
wetlands loss in the United States. National Wetlands Newsletter
6(5):3-6.

United Nations Environment Programme (UNEP). 1986. Report of the
International Conference on the Assessment of the Role of Carbon
Dioxide and of Other Greenhouse Gases in Climate Variations and
Associated Impacts. World Climate Programme, Villach, Austria,
October 9-15, 1985. Geneva, Switzerland: World Meteorological
Organization.

Wilcoxen, P.J. 1986. Coastal erosion and sea level rise: Implications
for ocean beach and San Francisco's westside transport project.
Coastal Zone Management Journal 14(3):173-191.

Yohe, G.W. 1988. The cost of not holding back the sea. Report for the
U.S. EPA, 1988.

14

THREATS TO BIOLOGICAL DIVERSITY AS THE EARTH WARMS

Robert L. Peters

This paper will describe how global warming, acting synergistically with habitat destruction and other human-caused threats, is likely to cause a wave of extinctions over the next 50 or so years.

BIOLOGICAL DIVERSITY

In the last 10 years a new concept has come to organize the thoughts of those concerned with the conservation of living things (see Soule, 1985). This concept is that of biological diversity, which means the variety of biological organisms in the world, including animals, plants, fungi, and bacteria. Biological diversity includes not only the species themselves, like tigers, lions, and the greater Antillean nightjar (_Caprimulgus cubanensis_), but also the genetic variation within species that gives apples, for example, both green and red individuals. Additionally, biological diversity includes the different ecological functions these organisms perform, from fixing nitrogen in the soil to cleansing the water we drink.

The idea of biodiversity is the crystallization of an evolution in thought about how to conserve living things. Historically, protection measures focused on animal or plant species that were recognized as being important economic or sport resources--a species-focused approach to conservation. Although species efforts are still appropriate in a variety of circumstances, this view has since broadened as our understanding of the importance of even apparently insignificant organisms has grown. Conservation now focuses primarily on conserving entire interdependent webs of species--an ecosystem approach. We now know that not only are all the biological pieces of the planetary jigsaw puzzle interdependent, but also that human society itself, even in an age of gene splicing, engineered food crops, and modern medicine, is vastly more dependent than most people realize upon the natural diversity of living organisms.

The potential gifts hidden in the array of wild species are rich: There is the genetic variability found in wild strains of common food crops, essential for constant infusion of disease- and pest-resistant characteristics into global agriculture. There are hundreds of potential new food crops, many superior to ones presently in widespread use.

139

There are new medicines, insecticides, and materials for industry. In the varied fungi and bacteria are blueprints for genetic engineering. Possibly even more important, the existing natural ecosystems give us what are called ecosystem services: recycling of nutrients, restoration of tired soils, production of oxygen, and even the production of rain (Ehrlich and Mooney, 1983). These are just some of the necessary functions supplied by the natural world. Their value in any one year is incalculable: In dollars, a single plant species may be worth billions to agriculture. In betterment of the human condition, a single obscure plant like the rosy periwinkle of Madagascar may save thousands of lives through the gift of medicine.

Although our knowledge of most species is overshadowed by our ignorance--indeed, most of the possibly 30 million species of organisms have not even been named--by extrapolation from species that have been investigated we can safely say that a large number are valuable. For example, of some 1500 previously uninvestigated plant species assayed for pharmacological activity, at least 15 percent promised medicinal importance (Caufield, 1985). Further, although our knowledge of ecological interactions is limited, we do know that species, like symbiotic fungi, that at first glance seem insignificant may be vital to ecosystem functioning. Clearly humans would do well to save as many species as possible. Aldo Leopold said, "If the biota, in the course of aeons, has built something we like but do not understand, then who but a fool would discard seemingly useless parts?" (Leopold, 1953).

Valuable though they are, a large percentage of biological resources are being destroyed by the increasing demands of growing human populations and economic development. Throughout the world, populations of most wild species are being decreased and fragmented, often to the degree that species of even present economic importance can no longer support viable industries, often to the degree that extinctions occur. For example, the blue pike (_Stizostedion vitreum glaucum_), a dominant fish of the U.S. Great Lakes, provided over 1 billion pounds of protein between 1888 and 1962, only to become extinct suddenly around 1965, most probably due to a combination of overfishing, pollution, and introduced species (Ono et al., 1983).

The best estimate of total world extinctions is still the estimate made by Lovejoy in 1980 for the _Global 2000_ report. He estimated, based on deforestation rates, that between 15 and 20 percent of all species on earth would become extinct by the year 2000 due to destruction of tropical forests alone (Lovejoy, 1980). This gives us a rough percentage rate, but to estimate absolute numbers of extinctions we need to know how many species exist on earth. Until the 1980s, estimates ranged between roughly 3 million and 10 million species. However, T. L. Erwin of the Smithsonian Institute has recently found that there are huge numbers of arthropod species, particularly beetles, high up in the Amazon rain forest canopy (1982, 1983). This has led to an upward revision of the world total to perhaps 30 million species. If this estimate is correct and given Lovejoy's projections, it means that between 1.5 million and 6 million species will be lost due to habitat destruction alone by the end of this century. This estimate is probably somewhat low because deforestation rates have increased substantially since Lovejoy made his estimates (see below).

Although population diminution or extinction of any particular species may hardly be noted, particularly if its contribution to human society is subtle or potential only, the cumulative effect of the processes of biological impoverishment may be easily seen in many parts of the world, from the arid Sahel region of Africa, where overgrazing and tree cutting have contributed to desertification and attendant human misery, to the United States, where a variety of species, including the California condor (<u>Gymnogyps californianus</u>), the largest bird of prey in North America, hover on the edge of extinction.

The primary cause of species loss today is habitat destruction. At least 71,000 km^2 of primary tropical forest were being cleared yearly throughout the world at the time of the last definitive estimate in 1982 (Lanly, 1982). However this rate is increasing rapidly, as at least 77,000 km^2 of virgin forest were cleared and burned in 1977 in the Brazilian Amazon alone (Simons, 1988). The result of such habitat destruction is to erase some species entirely and to leave others as remnant populations surviving in habitat fragments surrounded by cleared and hostile land.

Understanding the effects of this fragmentation is key to understanding how climate change will impact wild ecosystems in the disturbed landscape of the twentieth and twenty-first centuries. Fragmentation by definition makes populations smaller and isolated from each other, and small, isolated populations are at much greater risk from climate change than are larger ones. (Other factors that reduce populations, like pollution and hunting, will also make species more vulnerable to climate change, but habitat destruction will have the greatest effect.) Together, habitat destruction and global warming will act synergistically to cause many more extinctions than either will alone, as detailed below in the section on "Synergy of Habitat Destruction and Climate Change."

THE NATURE OF THE ECOLOGICALLY SIGNIFICANT CHANGES

There is widespread consensus that ecologically significant global warming will occur during the next century. For example, Hansen et al. (1988a) have said, "we can confidently state that major greenhouse climate changes are a certainty." It is expected that within the next 40 years greenhouse trace gases in the atmosphere, including carbon dioxide (CO_2), chlorofluorocarbons, and methane, will reach a concentration equivalent to double the preindustrial concentration of CO_2. The National Academy of Sciences and others have estimated that this concentration of greenhouse gases will be sufficient to raise the earth's temperature by $3 \pm 1.5°C$ (Hansen et al., 1988b; NRC, 1983, 1987; Schneider and Londer, 1984; WMO, 1982). More recent estimates suggest that warmings as high as $4.2 \pm 1.2°C$ (Schlesinger, 1989), or even 8 to 10°C (Lashof, 1989), cannot be ruled out. Because of a time lag caused by thermal inertia of the oceans, some of this warming will be delayed by 30 to 40 years beyond the time that a doubling equivalent of CO_2 is reached (EPA, 1988), but substantial warming could occur soon--the Goddard Institute for Space Studies (GISS) model projects a 2°C rise by

2020 A.D. (Rind, 1989). Such general circulation models have many uncertainties, but they provide the best estimates possible. As discussed below, this transitional warming would cause profound ecological change well before 3 or 4°C is reached--warming of less than 1°C would have substantial ecological effects.

It should be stressed that although projections can be made about global averages, regional projections are much less certain (Schneider, 1988). It is known that warming will not be even over the earth, with warming likely to be greater in the high latitudes, for example, than in the low latitudes (Hansen et al., 1988a). Regional and local peculiarities of typography and circulation will play a strong role in determining local climates.

For the purpose of discussion in this paper, I will take average global warming to be 3°C, since this is a commonly used benchmark, but it must be recognized that additional warming well beyond 3°C may be reached during the next century if the production of anthropogenic greenhouse gases continues. I will also make the conservative assumption that 3°C warming will not be reached until 2070 A.D. Additional warming or faster warming would cause additional biological disruption beyond that laid out in this paper.

The threats to natural systems are serious for the following reasons. First, 3°C of warming would present natural systems with a warmer world than has been experienced in the past 100,000 years (Schneider and Londer, 1984). An increase of 4°C would make the earth its warmest since the Eocene, 40 million years ago (Barron, 1985; see Webb, 1990). This warming would not only be large compared to recent natural fluctuations, but it would also occur very rapidly, perhaps 15 to 40 times faster than past natural changes (Gleick et al., 1990). For reasons discussed below, such a rate of change may exceed the ability of many species to adapt. Even widespread species are likely to have drastically curtailed ranges, at least in the short term. Moreover, human encroachment and habitat destruction will make wild populations of many species small and vulnerable to local climate changes.

Second, ecological stress would not be caused by temperature rise alone. Changes in global temperature patterns would trigger widespread alterations in rainfall patterns (Hansen et al., 1981; Kellogg and Schware, 1981; Manabe et al., 1981), and we know that for many species precipitation is a more important determinant of survival than is temperature per se. Indeed, except at the tree line, rainfall is the primary determinant of vegetation structure, trees occurring only where annual precipitation is in excess of 300 mm (Woodward, 1990). Because of global warming, some regions would see dramatic increases in rainfall, and others would lose their present vegetation because of drought. For example, the U.S. Environmental Protection Agency (EPA, 1988) concluded, based on several studies, that a long-term drying trend is likely in the midlatitude, interior continental regions during the summer. Specifically, based on rainfall patterns during past warming periods, Kellogg and Schware (1981) projected that substantial decreases in rainfall in North America's Great Plains are possible--perhaps as much as 40 percent by the early decades of the next century.

Other environmental factors important in determining vegetation type and health would change because of global warming: Soil chemistry would change (Kellison and Weir, 1987), as, for example, changes in storm patterns alter leaching and erosion rates (Harte et al., 1990). Increased CO_2 concentrations may accelerate the growth of some plants at the expense of others (NRC, 1983; Strain and Bazzaz, 1983), possibly destabilizing natural ecosystems. And rises in sea level may inundate coastal biological communities (NRC, 1983; Hansen et al., 1981; Hoffman et al., 1983; Titus et al., 1984).

As mentioned, it is generally concluded from a variety of computer projections that warming will be relatively greater at higher latitudes (Hansen et al., 1987). This suggests that, although tropical systems may be more diverse and are currently under great threat because of habitat destruction, temperate zone and arctic species may ultimately be in greater jeopardy from climate change. Arctic vegetation would experience widespread changes (Edlund, 1987). A recent attempt to map climate-induced changes in world biotic communities projected that high-latitude communities would be particularly stressed (Emanuel et al., 1985), and boreal forest, for example, was projected to decrease by 37 percent in response to global warming of 3°C.

A final point, important in understanding species response to climate change, is that weather is variable, and extreme events, like droughts, floods, blizzards, and hot or cold spells, may have greater effects on species distributions than does average climate per se (e.g., Knopf and Sedgwick, 1987). For example, in northwestern forests global warming is expected to increase the frequency of fires, leading to rapid alteration of forest character (Franklin, 1990).

SPECIES' RANGES SHIFT IN RESPONSE TO CLIMATE CHANGE

We know that when temperature and rainfall patterns change, species' ranges change. Not surprisingly, species tend to track their climatic optima, retracting their ranges where conditions become unsuitable while expanding them where conditions improve (Peters and Darling, 1985; Ford, 1982). Even very small temperature changes of less than 1°C within this century have been observed to cause substantial range changes. For example, the white admiral butterfly (Ladoga camilla) and the comma butterfly (Polygonia c-album) greatly expanded their ranges in the British Isles during the past century as the climate warmed approximately 0.5°C (see Ford, 1982). The birch (Betula pubescens) responded rapidly to warming during the first half of this century by expanding its range north into the Swedish tundra (Kullman, 1983).

On a larger ecological and temporal scale, entire vegetation types have shifted in response to past temperature changes no larger than those that may occur during the next 100 years or less (Baker 1983; Bernabo and Webb, 1977; Butzer, 1980; Flohn, 1979; Muller, 1979; Van Devender and Spaulding, 1979). As the earth warms, species tend to shift to higher latitudes and altitudes. From a simplified point of view, rising temperatures have caused species to colonize new habitats toward the poles, often while their ranges contracted away from the equator as conditions there became unsuitable.

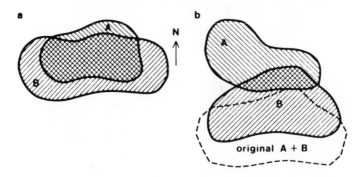

FIGURE 14.1 (a) Initial distribution of two species, A and B, whose ranges largely overlap. (b) In response to climate change, latitudinal shifting occurs at species-specific rates, and the ranges disassociate.

During several Pleistocene interglacials the temperature in North America was apparently 2 to 3°C higher than it is now. Sweet gum trees (<u>Liquidambar</u>) grew in southern Ontario (Wright, 1971); Osage oranges (<u>Maclura</u>) and papaws (<u>Asimina</u>) grew near Toronto, several hundred kilometers north of their present distributions; manatees swam in New Jersey; and tapirs and peccaries foraged in North Carolina (Dorf, 1976). During the last of these interglacials, which ended more than 100,000 years ago, vegetation in northwestern Europe, which is now boreal, was predominantly temperate (in Critchfield, 1980). Other significant changes in species' ranges have been caused by altered precipitation accompanying past global warming, including expansion of prairie in the American Midwest during a global warming episode approximately 7,000 years ago (Bernabo and Webb, 1977).

It should not be imagined, because species tend to shift in the same general direction, that existing biological communities move in synchrony. Conversely, because species shift at different rates in response to climate change, communities often disassociate into their component species (Figure 14.1). Recent studies of fossil packrat (<u>Neotoma</u> spp.) middens in the southwestern United States show that during the wetter, moderate climate of 22,000 to 12,000 years ago, there was not a concerted shift of plant communities. Instead, species responded individually to climatic change, forming stable but, by present-day standards, unusual assemblages of plants and animals (Van Devender and Spaulding, 1979). In eastern North America, too, post-glacial communities were often ephemeral associations of species, changing as individual ranges changed (Davis, 1983; Graham, 1986).

A final aspect of species response is that species may shift altitudinally as well as latitudinally. When climate warms, species shift upward. Generally, a short climb in altitude corresponds to a major shift in latitude: The 3°C cooling of a 500-m rise in elevation equals roughly the cooling achieved by a 250-km northward shift in latitude (MacArthur, 1972). Thus, during the middle Holocene, when temperatures in eastern North America were 2°C warmer than they are at

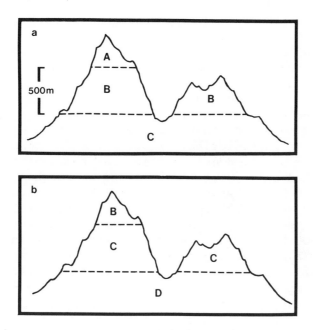

FIGURE 14.2 (a) Present altitudinal distribution of three species, A,
B, and C. (b) Species distribution after a 500-m shift in altitude in
response to a 3°C rise in temperature (based on Hopkin's bioclimatic
law; MacArthur, 1972). Species A becomes locally extinct. Species B
shifts upward and the total area it occupies decreases. Species C
becomes fragmented and restricted to a smaller area, while species D
successfully colonizes the lowest-altitude habitats.

present, hemlock (Tsuga canadensis) and white pine (Pinus strobus) were
found 350 m higher on mountains than they are today (Davis, 1983).

Because mountain peaks are smaller than bases, as species shift
upward in response to warming, they typically occupy smaller and smaller
areas, have smaller populations, and may thus become more vulnerable to
genetic and environmental pressures (see Murphy and Weiss, 1990).
Species originally situated near mountaintops may have no habitat to
move up to and may be entirely replaced by the relatively thermophilous
species moving up from below (Figure 14.2). Examples of past extinctions attributed to upward shifting include alpine plants once living on
mountains in Central and South America, where vegetation zones have
shifted upward by 1000 to 1500 m since the last glacial maximum (Flenley, 1979; Heusser, 1974).

MAGNITUDE OF PROJECTED LATITUDINAL SHIFTS

If the proposed CO_2-induced warming occurs, species shifts similar
to those in the Pleistocene will occur, and vegetation belts will move
hundreds of kilometers toward the poles (Davis and Zabinski, 1990; Frye,
1983; Peters and Darling, 1985). A 300-km shift in the temperate zone
is a reasonable minimum estimate for a 3°C warming, based on the

positions of vegetation zones during analogous warming periods in the past (Dorf, 1976; Furley et al., 1983).

Additional confirmation that shifts of this magnitude or greater may occur comes from attempts to project future range shifts for some species by looking at their ecological requirements. For example, the forest industry is concerned about the future of commercially valuable North American species, like the loblolly pine (Pinus taeda L.). This species is limited on its southern border by moisture stress on seedlings. Based on its physiological requirements for temperature and moisture, Miller et al. (1987) projected that the southern range limit of the species would shift approximately 350 km northward in response to a global warming of 3°C. Davis and Zabinski (1990) have projected possible northward range withdrawals among several North American tree species, including sugar maple (Acer saccharum) and beech (Fagus grandifolia), of from 600 km to possibly as much as 2000 km in response to the warming caused by a doubled CO_2 concentration. Beech would be most responsive, withdrawing from its present southern extent along the Gulf Coast and retreating into Canada.

MECHANISMS UNDERLYING RANGE SHIFTS

The range shifts described above are the sum of many local processes of extinction and colonization that occur in response to climate-caused changes in suitability of habitats. These changes in habitat suitability are determined by both direct climate effects on physiology, including temperature and precipitation, and indirect effects secondarily caused by other species, themselves affected by temperature.

There are numerous examples of climate directly influencing survival and thereby distribution. In animals, the direct range-limiting effects of excessive warmth include lethality, as in corals (Glynn, 1984), and interference with reproduction, as in the large blue butterfly, Maculinea arion (Ford, 1982). In plants, excessive heat and associated decreases in soil moisture may decrease survival and reproduction. Coniferous seedlings, for example, are injured by soil temperatures over 45°C, although other types of plants can tolerate much higher temperatures (see Daubenmire, 1962). Many plants have their northern limits determined by minimum temperature isotherms below which some key physiological process does not occur. For instance, the grey hair grass (Corynephorus canescens) is largely unsuccessful at germinating seeds below 15°C and is bounded to the north by the 15°C July mean isotherm (Marshall, 1978). Moisture extremes exceeding physiological tolerances also determine species' distributions. Thus, the European range of the beech tree (Fagus sylvatica) ends to the south where rainfall is less than 600 mm annually (Seddon, 1971), and dog's mercury (Mercurialis perennis), an herb restricted to well-drained sites in Britain, cannot survive in soil where the water table reaches as high as 10 cm below the soil surface (see Ford, 1982).

The physiological adaptations of most species to climate are conservative, and it is unlikely that most species could evolve significantly new tolerances in the time allotted to them by the coming warming trend.

Indeed, the evolutionary conservatism in thermal tolerance of many plant and animal species--beetles, for example (Coope, 1977)--is the underlying assumption that allows us to infer past climates from faunal and plant assemblages.

Interspecific interactions altered by climate change will have a major role in determining new species distributions. Temperature can influence predation rates (Rand, 1964), parasitism (Aho et al., 1976), and competitive interactions (Beauchamp and Ullyott, 1932). Changes in the ranges of tree pathogens and parasites may be important in determining future tree distributions (Winget, 1988). Soil moisture is a critical factor in mediating competitive interactions among plants, as is the case where the dog's mercury (Mercurialis perennis) excludes oxlip (Primula elatior) from dry sites (Ford, 1982).

Given the new associations of species that occur as climate changes, many species will face "exotic" competitors for the first time. Local extinctions may occur as climate change causes increased frequencies of droughts and fires, favoring invading species. One species that might spread, given such conditions, is Melaleuca quinquenervia, a bamboo-like Australian eucalypt. This species has already invaded the Florida Everglades, forming dense monotypic stands where drainage and frequent fires have dried the natural marsh community (Courtenay, 1978; Myers, 1983).

The preceding effects, both direct and indirect, may act in synergy, as when drought makes a tree more susceptible to insect parasitism.

DISPERSAL RATES AND BARRIERS

The ability of species to adapt to changing conditions will depend to a large extent on their ability to track shifting climatic optima by dispersing colonists. In the case of warming, a North American species, for example, would most likely need to establish colonies to the north or at higher elevations. Survival of plant and animal species would therefore depend either on long-distance dispersal of colonists, such as seeds or migrating animals, or on rapid iterative colonization of nearby habitat until long-distance shifting results. A plant's intrinsic ability to colonize will depend on its ecological characteristics, including fecundity, viability and growth characteristics of seeds, nature of the dispersal mechanism, and ability to tolerate selfing and inbreeding upon colonization. If a species' intrinsic colonization ability is low, or if barriers to dispersal are present, extinction may result if all of its present habitat becomes unsuitable.

There are many cases where complete or local extinction has occurred because species were unable to disperse rapidly enough when the climate changed. For example, a large, diverse group of plant genera, including water-shield (Brassenia), sweet gum (Liquidambar), tulip tree (Liriodendron), magnolia (Magnolia), moonseed (Menispermum), hemlock (Tsuga), arbor vitae (Thuja), and white cedar (Chamaecyparis), had a circumpolar distribution in the Tertiary (Tralau, 1973). But during the Pleistocene ice ages, all became extinct in Europe while surviving in North America. Presumably, the east-west orientation of such barriers as the Pyrenees, the Alps, and the Mediterranean, which blocked southward migration, was partly responsible for their extinction (Tralau, 1973).

Other species of plants and animals thrived in Europe during the cold periods but could not survive conditions in postglacial forests. One such previously widespread dung beetle, Aphodius hodereri, is now extinct throughout the world except in the high Tibetan plateau where conditions remain cold enough for its survival (Cox and Moore, 1985). Other species, like the Norwegian mugwort (Artemisia novegica) and the springtail Tetracanthella arctica, now live primarily in the boreal zone but also survive in a few cold, mountaintop refugia in temperate Europe (Cox and Moore, 1985).

These natural changes were slow compared to changes predicted for the near future. The change to warmer conditions at the end of the last ice age spanned several thousand years yet is considered rapid by geologic standards (Davis, 1983). We can deduce that, if such a slow change was too fast to allow many species to adapt, the projected warming--possibly 40 times faster--will have more severe consequences. For widespread, abundant species, like the loblolly pine (modeled by Miller et al., 1987), even substantial range retraction might not threaten extinction; but rare, localized species, whose entire ranges might become unsuitable, would be threatened unless dispersal and colonization were successful. Even for widespread species, major loss of important ecotypes and associated germplasm is likely (see Davis and Zabinski, 1990).

A key question is whether the dispersal capabilities of most species prepare them to cope with the coming rapid warming. If the climatic optima of temperate zone species do shift hundreds of kilometers toward the poles within the next 100 years, then these species will have to colonize new areas rapidly. To survive, a localized species whose present range becomes unsuitable might have to shift poleward at several hundred kilometers or faster per century. Although some species, such as plants propagated by spores or "dust" seeds, may be able to match these rates (Perring, 1965), many species could not disperse quickly enough to compensate for the expected climatic change without human assistance (see Rapoport, 1982), particularly given the presence of dispersal barriers. Even wind-assisted dispersal may fall short of the mark for many species. In the case of the Engelmann spruce (Picea engelmannii), a tree with light, wind-dispersed seeds, fewer than 5 percent of seeds travel even 200 m downwind, leading to an estimated migration rate of 1 to 20 km per century (Seddon, 1971); this reconciles well with rates derived from fossil evidence for North American trees of between 10 and 45 km per century (Davis and Zabinski, 1990; Roberts, 1989). As described in the next section, many migration routes will likely be blocked by the cities, roads, and fields replacing natural habitat.

Although many animals may be, in theory, highly mobile, the distribution of some is limited by the distributions of particular plants, i.e., suitable habitat; their dispersal rates may therefore be determined largely by those of co-occurring plants. Behavior may also restrict dispersal even of animals physically capable of large movements. Dispersal rates below 2.0 km per year have been measured for several

species of deer (Rapoport, 1982), and many tropical deep-forest birds
simply do not cross even very small unforested areas (Diamond, 1975).
On the other hand, some highly mobile animals may shift rapidly, as have
some European birds (see Edgell, 1984).

Even if animals can disperse efficiently, suitable habitat may be
reduced under changing climatic conditions. For example, it has been
suggested that tundra nesting habitat for migratory shore birds might be
reduced by high-arctic warming (Myers, 1988).

SYNERGY OF HABITAT DESTRUCTION AND CLIMATE CHANGE

We know that even slow, natural climate change caused species to
become extinct. What is likely to happen given the environmental conditions of the coming century?

Some clear implications for conservation follow from the preceding
discussion of dispersal rates. Any factor that would decrease the probability that a species could successfully colonize new habitat would
increase the probability of extinction. Thus, as previously described,
species are more likely to become extinct if there are physical barriers
to colonization, such as oceans, mountains, and cities. Further, species
are more likely to become extinct if their remaining populations are
small. Smaller populations mean that fewer colonists can be sent out and
that the probability of successful colonization is smaller.

Species are more likely to become extinct if they occupy a small
geographic range, which is less likely than a larger range to offer some
suitable remaining refuge when climate changes. Also, if a species has
lost much of its range because of some other factor, like clearing of
the richer and moister soils for agriculture, it is possible that remaining populations are located in poor habitat and are therefore more
susceptible to new stresses.

For many species, all of these conditions will be met by human-
caused habitat destruction, which increasingly confines the natural
biota to small patches of original habitat, patches isolated by vast
areas of human-dominated urban or agricultural lands.

Habitat destruction in conjunction with climate change sets the
stage for an even larger wave of extinction than that previously
imagined, based on consideration of human encroachment alone. Small,
remnant populations of most species, surrounded by cities, roads,
reservoirs, and farmland, will have little chance of reaching new
habitat if climate change makes the old unsuitable. Few animals or
plants would be able to cross Los Angeles on the way to new habitat.
Figure 14.3 illustrates the combined effects of habitat loss and warming
on a hypothetical reserve.

AMELIORATION AND MITIGATION

Because of the difficulty of predicting regional and local changes,
conservationists and reserve managers must deal with increased uncertainty in making long-range plans. However, even given imprecise

150

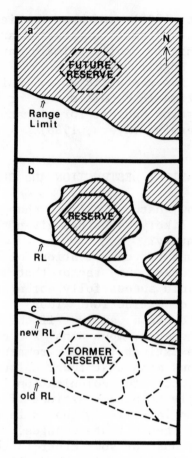

FIGURE 14.3 Climatic warming may cause species within biological
reserves to disappear. Hatching indicates: (a) species distribution
before either human habitation or climate change (range limit, RL,
indicates southern limit of species range); (b) fragmented species
distribution after human habitation but before climate change; (c)
species distribution after human habitation and climate change.

regional projections, informed guesses can be made at least about the
general direction of change, specifically that most areas will tend to be
hotter and that continental interiors in particular are likely to
experience decreased soil moisture.

How might the threats posed by climatic change to natural communities be mitigated? One basic truth is that the less populations are
reduced by development now, the more resilient they will be to climate
change. Thus, sound conservation now, in which we try to conserve more
than just the minimum number of individuals of a species necessary for
present survival, would be an excellent way to start planning for climate
change.

In terms of responses specifically directed at the effects of
climate change, the most environmentally conservative response would be

to halt or slow global warming by cutting back on production of fossil fuels, methane, and chlorofluorocarbons. Extensive planting of trees to capture CO_2 could help slow the rise in CO_2 concentrations (Sedjo, 1989). Nonetheless, even were the production of all greenhouse gases stopped today, it is very likely that there are now high enough concentrations in the air to cause ecologically significant warming after a brief lag period (Rind, 1989). Therefore, those concerned with the conservation of biological diversity must begin to plan mitigation activities.

To make intelligent plans for siting and managing reserves, we need more knowledge. We must refine our ability to predict future conditions in reserves. We also need to know more about how temperature, precipitation, CO_2 concentrations, and interspecific interactions determine range limits (e.g., Picton, 1984; Randall, 1982) and, most important, how they can cause local extinctions.

Reserves that suffer from the stresses of altered climatic regimes will require carefully planned and increasingly intensive management to minimize species loss. For example, modifying conditions within reserves may be necessary to preserve some species; depending on changes in moisture patterns, irrigation or drainage may be needed. Because of changes in interspecific interactions, competitors and predators may need to be controlled and invading species weeded out. The goal would be to maintain suitable conditions for desired species or species assemblages, much as the habitat of Kirtland's warbler is periodically burned to maintain pine woods (Leopold, 1978).

In attempting to understand how climatically stressed communities may respond and how they might be managed to prevent the gradual depauperization of their constituents, restoration studies, or more properly, "community creation" experiments can help. Communities may be created outside their normal climatic ranges to mimic the effects of climate change. One such "out-of-place" community is the Leopold Pines at the University of Wisconsin Arboretum in Madison, where there is periodically less rainfall than in the normal pine range several hundred kilometers to the north (Jordan, 1988). Researchers have found that, although the pines themselves do fairly well once established at the Madison site, many of the other species that would normally occur in a pine forest, especially the various herbs and small shrubs, have not flourished, despite several attempts to introduce them (Anderson et al., 1969).

If management measures are unsuccessful and old reserves do not retain necessary thermal or moisture characteristics, individuals of disappearing species might be transferred to new reserves. For example, warmth-intolerant ecotypes or subspecies might be transplanted to reserves nearer the poles. Other species may have to be periodically reintroduced in reserves that experience occasional climate extremes severe enough to cause extinction but where the climate would ordinarily allow the species to survive with minimal management. Such transplantations and reintroductions, particularly involving complexes of species, will often be difficult, but some applicable technologies are being developed (Botkin, 1977; Lovejoy, 1985).

To the extent that we can still establish reserves, pertinent information about changing climate and subsequent ecological response should be used in deciding how to design and locate them to minimize the effects of changing temperature and moisture.

o The existence of multiple reserves for a given species or community type increases the probability that, if one reserve becomes unsuitable for climatic reasons, the organisms may still be represented in another reserve.

o Reserves should be heterogeneous with respect to topography and soil types, so that even given climatic change, remnant populations may be able to survive in suitable microclimatic areas. Species may survive better in reserves with wide variations in altitude, since, from a climatic point of view, a small altitudinal shift corresponds to a large latitudinal one. Thus, to compensate for a 2°C rise in temperature, a northern hemisphere species can achieve almost the same result by increasing its altitude only some 500 m as it would by moving 300 km to the north (MacArthur, 1972).

o Corridors between reserves, important for other conservation reasons, would allow some natural migration of species to track climate shifting. Corridors along altitudinal gradients are likely to be most practical because they can be relatively short compared with the longer distances necessary to accommodate latitudinal shifting.

o As climatic models become more refined, pertinent information should be taken into consideration in making decisions about where to site reserves in order to minimize the effects of temperature and moisture changes. In the northern hemisphere, for example, where a northward shift in climatic zones is likely, it makes sense to locate reserves as near the northern limit of a species' or community's range as possible, rather than farther south, where conditions are likely to become unsuitable more rapidly.

o Maximizing the size of reserves will increase long-term persistence of species by increasing the probability that suitable microclimates exist, by increasing the probability of altitudinal variation, and by increasing the latitudinal distance available to shifting populations.

o Flexible zoning around reserves may allow us to actually move reserves in the future to track climatic optima, as, for example, by trading present rangeland for reserve land. The success of this strategy, however, would depend on a highly developed restoration technology capable of guaranteeing, in effect, the portability of species and whole communities.

CONCLUSION

The best solutions to the ecological upheaval resulting from climatic change are not yet clear. In fact, little attention has been paid to the problem. What is clear, however, is that these climatological changes would have tremendous impact on communities and populations isolated by development and by the middle of the next

153

century might dwarf any other consideration in planning for reserve management. The problem may seem overwhelming. One thing, however, is worth keeping in mind: The more fragmented and the smaller populations of species are, the less resilient they will be to the new stresses brought about by climate change. Thus, one of the best things that can be done in the short-term is to minimize further encroachment of development on existing natural ecosystems. Further, we must refine our climatological predictions and increase our understanding of how climate affects species, both individually and in their interactions with each other. Such studies may allow us to identify those areas where communities will be most stressed, as well as alternate areas where they might best be saved. Meanwhile, efforts to improve techniques for managing communities and ecosystems under stress, and also for restoring them when necessary, must be carried forward energetically.

ACKNOWLEDGMENTS

The bulk of the text on global warming was previously published as an article in Forest Ecology and Management, in a special 1989 volume containing the proceedings of the symposium on Conservation of Diversity in Forest Ecosystems, University of California at Davis, July 25, 1988. It draws heavily on other previously published versions, including those in Endangered Species Update (Vol. 5, No. 7, pp. 1-8) and Preparing for Climate Change: Proceedings of the First North American Conference on Preparing for Climate Change: A Cooperative Approach (J.C. Topping, Jr., ed., Government Institutes, Washington, D.C.). Many of the ideas and the three figures derive from Peters and Darling (1985); please see that paper also for a complete list of acknowledgments for help with this work.

REFERENCES

Aho, J.M., J.W. Gibbons, G.W. Esch. 1976. Relationship between thermal loading and parasitism in the mosquitofish. Pp. 213-218 in G.W. Esch and R.W. McFarlane, eds. Thermal Ecology II. Technical Information Center, Energy Research and Development Administration, Springfield, Va.

Anderson, R.C., O.L. Loucks, and A.M. Swain. 1969. Herbaceous response to canopy cover, light intensity and throughfall precipitation in coniferous forests. Ecology 50:255-263.

Baker, R.G. 1983. Holocene vegetational history of the western United States. Pp. 109-125 in H.E. Wright, Jr., ed. Late-Quaternary Environments of the United States. Volume 2. The Holocene. University of Minnesota Press, Minneapolis.

Barron, E.J. 1985. Explanations of the Tertiary global cooling trend. Palaeogeography, Palaeoclimatology, Palaeoecology 50:17-40.

Beauchamp, R.S.A., and P. Ullyott. 1932. Competitive relationships between certain species of fresh-water triclads. J. Ecol. 20:200-208.

154

Bernabo, J.C., and T. Webb III. 1977. Changing patterns in the Holocene pollen record of northeastern North America: a mapped summary. Quat. Res. 8:64-96.

Botkin, D.B. 1977. Strategies for the reintroduction of species into damaged ecosystems. Pp. 241-260 in J. Cairns, Jr., K.L. Dickson, and E.E. Herricks, eds. Recovery and Restoration of Damaged Ecosystems. University Press of Virginia, Charlottesville.

Butzer, K.W. 1980. Adaptation to global environmental change. Prof. Geogr. 32:269-278.

Caufield, C. 1985. In the Rainforest. Knopf, New York.

Coope, G.R. 1977. Fossil coleopteran assemblages as sensitive indicators of climatic changes during the Devensian (Last) cold stage. Philos. Trans. R. Soc. Lond. B. 280:313-340.

Courtenay, Jr., W.R. 1978. The introduction of exotic organisms. Pp. 237-252 in H.P. Brokaw, ed. Wildlife and America. Council on Environmental Quality. U.S. Government Printing Office, Washington, D.C.

Cox, B.C., and P.D. Moore. 1985. Biogeography: An Ecological and Evolutionary Approach. Blackwell Scientific Publications, Oxford.

Critchfield, W.B. 1980. Origins of the eastern deciduous forest. Pp. 1-14 in Proceedings, Dendrology in the Eastern Deciduous Forest Biome, September 11-13, 1979. Virginia Polytechnic Institute and State University School of Forestry and Wildlife Resources. Publ. FWS-2-80.

Daubenmire, R.F. 1962. Plants and Environment: A Textbook of Plant Autecology. John Wiley & Sons, New York.

Davis, M.B. 1983. Holocene vegetational history of the eastern United States. Pp. 166-181 in H.E. Wright, Jr., ed. Late-Quaternary Environments of the United States. Volume 2. The Holocene. University of Minnesota Press, Minneapolis.

Davis, M.B., and C. Zabinski. 1990. Changes in geographical range resulting from greenhouse warming effects on biodiversity in forests. In R.L. Peters and T.E. Lovejoy, eds. Proceedings of World Wildlife Fund's Conference on Consequences of Global Warming for Biological Diversity. Yale University Press, New Haven, Conn., forthcoming.

Diamond, J.M. 1975. The island dilemma: lessons of modern biogeographic studies for the design of natural preserves. Biol. Conserv. 7:129-146.

Dorf, E. 1976. Climatic changes of the past and present. Pp. 384-412 in C.A. Ross, ed. Paleobiogeography: Benchmark Papers in Geology 31. Dowden, Hutchinson, and Ross, Stroudsburg, Pa.

Edgell, M.C.R. 1984. Trans-hemispheric movements of Holarctic Anatidae: the Eurasian wigeon (Anas penelope L.) in North America. J. Biogeogr. 11:27-39.

Edlund, S.A. 1987. Effects of climate change on diversity of vegetation in arctic Canada. Pp. 186-193 in J.C. Topping, Jr., ed. Preparing for Climate Change: Proceedings of the First North American Conference on Preparing for Climate Change: A Cooperative Approach. Government Institutes, Washington, D.C.

Ehrlich, P.R., and H.A. Mooney. 1983. Extinction, substitution, and ecosystem services. Bioscience 33(4):248-254.

Emanuel, W.R., H.H. Shugart, and M.P. Stevenson. 1985. Response to comment: "Climatic change and the broad-scale distribution of terrestrial ecosystem complexes." Clim. Change 7:457-460.

Environmental Protection Agency (EPA). 1988. The Potential Effects of Global Climate Change on the United States: Draft Report to Congress, Vol. 1. EPA, Washington, D.C.

Erwin, T.L. 1982. Tropical forests: their richness in Coleoptera and other arthropod species. Coleopt. Bull. 36:74-75.

Erwin, T.L. 1983. Beetles and other insects of tropical forest canopies at Manaus, Brazil, sampled by insecticidal fogging. Pp. 59-75 in T.C. Whitmore and A.C. Chadwick, eds. Tropical Rain Forest Ecology and Management. Blackwell Scientific Publishers, Oxford.

Flenley, J.R. 1979. The Equatorial Rain Forest. Butterworths, London.

Flohn, H. 1979. Can climate history repeat itself? Possible climatic warming and the case of paleoclimatic warm phases. Pp. 15-28 in W. Bach, J. Pankrath, and W.W. Kellogg, eds. Man's Impact on Climate. Elsevier Scientific Publishing, Amsterdam.

Ford, M.J. 1982. The Changing Climate. George Allen and Unwin, London.

Franklin, J.F. 1990. Effects of global climatic change on forests in North Western North America. In R.L. Peters and T.E. Lovejoy, eds. Proceedings of the Conference on Consequences of Global Warming for Biological Diversity. Yale University Press, New Haven, Conn., forthcoming.

Frye, R. 1983. Climatic change and fisheries management. Nat. Resour. J. 23:77-96.

Furley, P.A., W.W. Newey, R.P. Kirby, and J.McG. Hotson. 1983. Geography of the Biosphere. Butterworths, London.

Gleick, P.H., L. Mearns, and S.H. Schneider. 1990. Climate-change scenarios for impact assessment. In R.L. Peters and T.E. Lovejoy, eds. Proceedings of World Wildlife Fund's Conference on Consequences of the Greenhouse Effect for Biological Diversity. Yale University Press, New Haven, Conn., forthcoming.

Glynn, P. 1984. Widespread coral mortality and the 1982-83 El Nino warming event. Environ. Conserv. 11(2):133-146.

Graham, R.W. 1986. Plant-animal interactions and Pleistocene Extinctions. Pp. 131-154 in D.K. Elliott, ed. Dynamics of Extinction. Wiley & Sons, Somerset, N.J.

Hansen, J., D. Johnson, A. Lacis, S. Lebedeff, P. Lee, D. Rind, and G. Russell. 1981. Climate impact of increasing atmospheric carbon dioxide. Science 213:957-966.

Hansen, J., A. Lacis, D. Rind, G. Russell, I. Fung, and S. Lebedeff. 1987. Evidence for future warming: how large and when. In W.E. Shands and J.S. Hoffman, eds. The Greenhouse Effect, Climate Change, and U.S. Forests. Conservation Foundation, Washington, D.C.

Hansen, J., I. Fung, A. Lacis, S. Lebedeff, D. Rind, R. Ruedy, G. Russell. 1988a. Prediction of near-term climate evolution: what can we tell decision-makers now? Pp. 35-47 in Preparing for Climate Change: Proceedings of the First North American Conference on

Preparing for Climate Change: A Cooperative Approach. Government Institutes, Washington, D.C.

Hansen, J., I. Fung, A. Lacis, D. Rind, S. Lebedeff, R. Ruedy, and G. Russell. 1988b. Global climate changes as forecast by Goddard Institute for Space Studies three-dimensional model. J. Geophys. Res. 93(D8):9341-9364.

Harte, J., M. Torn, and D. Jensen. 1990. The nature and consequences of indirect linkages between climate change and biological diversity. In R.L. Peters and T.E. Lovejoy, eds. Proceedings of World Wildlife Fund's Conference on Consequences of Global Warming for Biological Diversity. Yale University Press, New Haven, Conn., forthcoming.

Heusser, C.J. 1974. Vegetation and climate of the southern Chilean lake district during and since the last interglaciation. Quat. Res. 4:290-315.

Hoffman, J.S., D. Keyes, and J.G. Titus. 1983. Projecting Future Sea Level Rise. U.S. Environmental Protection Agency, Washington, D.C.

Jordan III, W.R. 1988. Ecological restoration: Reflections on a half century of experience at the University of Wisconsin-Madison Arboretum. Pp. 311-316 in E.O. Wilson, ed. Biodiversity. National Academy Press, Washington, D.C.

Kellison, R.C., and R.J. Weir. 1987. Selection and breeding strategies in tree improvement programs for elevated atmospheric carbon dioxide levels. In W.E. Shands and J.S. Hoffman, eds. The Greenhouse Effect, Climate Change, and U.S. Forests. Conservation Foundation, Washington, D.C.

Kellogg, W.W., and R. Schware. 1981. Climate Change and Society: Consequences of Increasing Atmospheric Carbon Dioxide. Westview Press, Boulder, Colo.

Knopf, F.L., and J.A. Sedgwick. 1987. Latent population responses of summer birds to a catastrophic, climatological event. The Condor 89:869-873.

Kullman, L. 1983. Past and present tree lines of different species in the Handolan Valley, Central Sweden. Pp. 25-42 in P. Morisset and S. Payette, eds. Tree Line Ecology. Centre d'etudes nordiques de l'Universite Laval, Quebec.

Lanly, J. 1982. Tropical Forest Resources, FAO Forestry Paper Number 30. Food and Agriculture Organization of the United Nations, Rome.

Lashof, D.A. 1989. The dynamic greenhouse: feedback processes that may influence future concentrations of atmospheric trace gases and climatic change. Climatic Change 14:213-242.

Leopold, A. 1953. Round River. Oxford University Press, New York.

Leopold, A.S. 1978. Wildlife and forest practice. Pp. 108-120 in H.P. Brokaw, ed. Wildlife and America. Council on Environmental Quality, U.S. Government Printing Office, Washington, D.C.

Lovejoy, T.E. 1980. A projection of species extinctions. Pp. 328-331 in The Global 2000 Report to the President: Entering the Twenty-First Century. Council on Environmental Quality and the Department of State. U.S. Government Printing Office, Washington, D.C.

Lovejoy, T.E. 1985. Rehabilitation of degraded tropical rainforest lands. Commission on Ecology Occasional Paper 5. International

Union for the Conservation of Nature and Natural Resources, Gland, Switzerland.

MacArthur, R.H. 1972. Geographical Ecology. Harper & Row, New York.

Manabe, S., R.T. Wetherald, and R.J. Stouffer. 1981. Summer dryness due to an increase of atmospheric CO_2 concentration. Clim. Change 3:347-386.

Marshall, J.K. 1978. Factors limiting the survival of Corynephorus canescens (L) Beauv. in Great Britain at the northern edge of its distribution. Oikos 19:206-216.

Miller, W.F., P.M. Dougherty, and G.L. Switzer. 1987. Rising CO_2 and changing climate: major southern forest management implications. The Greenhouse Effect, Climate Change, and U.S. Forests. Conservation Foundation, Washington, D.C.

Muller, H. 1979. Climatic changes during the last three interglacials. Pp. 29-41 in W. Bach, J. Pankrath, and W.W. Kellogg, eds. Man's Impact on Climate. Elsevier Scientific Publishing, Amsterdam.

Murphy, D.D., and S.B. Weiss. 1990. The effects of climate change on biological diversity in western North America: species losses and mechanisms. In R.L. Peters and T.E. Lovejoy, eds. Proceedings of World Wildlife Fund's Conference on the Consequences of the Greenhouse Effect for Biological Diversity. Yale University Press, New Haven, Conn., forthcoming.

Myers, J.P. 1988. The likely impact of climate change on migratory birds in the arctic. Presentation at Seminar on Impact of Climate Change on Wildlife, January 21-22, 1988. Climate Institute, Washington, D.C.

Myers, R.L. 1983. Site susceptibility to invasion by the exotic tree Melaleuca quinquenervia in southern Florida. J. Appl. Ecol. 20(2):645-658.

National Research Council (NRC). 1983. Changing Climate. National Academy Press, Washington, D.C.

National Research Council (NRC). 1987. Current Issues in Atmospheric Change. National Academy Press, Washington, D.C.

Ono, R.D, J.D. Williams, and A. Wagner. 1983. Vanishing Fishes of North America. Stone Wall Press, Washington, D.C.

Perring, F.H. 1965. The advance and retreat of the British flora. Pp. 51-59 in C.J. Johnson and L.P. Smith, eds. The Biological Significance of Climatic Changes in Britain. Academic Press, London.

Peters, R.L., and J.D. Darling. 1985. The greenhouse effect and nature reserves. BioScience 35(11):707-717.

Picton, H.D. 1984. Climate and the prediction of reproduction of three ungulate species. J. Appl. Ecol. 21:869-879.

Rand, A.S. 1964. Inverse relationship between temperature and shyness in the lizard Anolis lineatopus. Ecology 45:863-864.

Randall, M.G.M. 1982. The dynamics of an insect population throughout its altitudinal distribution: Coleophora alticolella (Lepidoptera) in northern England. J. Anim. Ecol. 51:993-1016.

Rapoport, E.H. 1982. Areography: Geographical Strategies of Species. Pergamon Press, New York.

Rind, D. 1989. A character sketch of greenhouse. EPA Journal 15(1): 4-7.

Roberts, L. 1989. How fast can trees migrate? Science 243:735-737.

Schlesinger, M.E. 1989. NATO Conference Proceedings, in press.

Schneider, S.H. 1988. The greenhouse effect: What we can or should do about it. Pp. 19-34 in Preparing for Climate Change: Proceedings of the First North American Conference on Preparing for Climate Change: A Cooperative Approach. Government Institutes, Washington, D.C.

Schneider, S.H., and R. Londer. 1984. The Coevolution of Climate and Life. Sierra Club Books, San Francisco.

Seddon, B. 1971. Introduction to Biogeography. Barnes and Noble, New York.

Sedjo, R.A. 1989. Forests: a tool to moderate global warming? Environment 31(1):14-20.

Simons, M. 1988. Vast Amazon fires, man-made, linked to global warming. New York Times, August, 11, 1988.

Soule, M.E. 1985. What is conservation biology? BioScience 35(11):727-734.

Strain, B.R., and F.A. Bazzaz. 1983. Terrestrial plant communities. Pp. 177-222 in E.R. Lemon, ed. CO_2 and Plants. Westview Press, Boulder, Colo.

Titus, J.G., T.R. Henderson, and J.M. Teal. 1984. Sea level rise and wetlands loss in the United States. National Wetlands Newsletter 6(5):3-6.

Tralau, H. 1973. Some quaternary plants. Pp. 499-503 in A. Hallam, ed. Atlas of Palaeobiogeography. Elsevier Scientific Publishing, Amsterdam.

Van Devender, T.R., and W.G. Spaulding. 1979. Development of vegetation and climate in the southwestern United States. Science 204:701-710.

Webb III, T. 1990. Past Changes in Vegetation and Climate: Lessons for the Future. In Peters, R.L. and T.E. Lovejoy, eds. Proceedings of World Wildlife Fund's Conference on Consequences of the Greenhouse Effect for Biological Diversity. Yale University Press, New Haven, Conn., forthcoming.

Winget, C.H. 1988. Forest management strategies to address climate change. Pp. 328-333 in Preparing for Climate Change: Proceedings of the First North American Conference on Preparing for Climate Change: A Cooperative Approach. Government Institutes, Washington, D.C.

Woodward, F.I. 1990. Review of the effects of climate on vegetation: ranges, competition and composition. In R.L. Peters and T.E. Lovejoy, eds. Proceedings of World Wildlife Fund's Conference on Consequences of the Greenhouse Effect for Biological Diversity. Yale University Press, New Haven, Conn., forthcoming.

World Meteorological Organization (WMO). 1982. Report of the JSC/CAS: A Meeting of Experts on Detection of Possible Climate Change (Moscow, October 1982). Geneva, Switzerland: Rep. WCP29.

Wright, Jr., H.E. 1971. Late Quaternary vegetational history of North America. Pp. 425-464 in K.K. Turekian, ed. The Late Cenozoic Glacial Ages. Yale University Press, New Haven, Conn.

15

DEFORESTATION AND ITS ROLE IN POSSIBLE
CHANGES IN THE BRAZILIAN AMAZON

Eneas Salati, Reynaldo Luiz Victoria,
Luiz Antonio Martinelli, and Jeffrey Edward Richey

Global deforestation, and in particular deforestation of the humid tropics, has become a controversial issue both scientifically and politically. The polemical aspects of the question are due primarily to the lack of detailed information on the changes that have occurred and on the processes that would permit the establishment of a sustainable use of the renewable natural resources. One of the most important global aspects of deforestation is the transfer of carbon stored in the forest biomass to the atmosphere (Woodwell et al., 1978; Houghton et al., 1985). In 1980 it was estimated that 25 percent of the global atmospheric carbon dioxide (CO_2) emissions was derived from the transformation of forests into annual cropland or grassland in the tropics (Houghton et al., 1987). Extinction of flora and fauna may be expected whenever the diversity of a natural forest ecosystem is reduced by the substitution of any other single species or less diverse ecosystem (Prance, 1986). Depending on the extent of the cleared area, deforestation can affect the climate at micro- and meso-scales, with local and regional and possibly global consequences (Salati and Marques, 1984; Henderson-Sellers, 1987).

Deforestation and its consequences in the Amazon basin are discussed in the present paper. The Amazon, the richest and most diverse ecosystem on the planet, is presently undergoing rapid transformations due to the expansion of new settlements undertaking diverse activities.

CAUSES FOR DEFORESTATION

In the beginning of the colonization of Brazil by Europeans in 1530, and in the centuries thereafter, deforestation in Amazônia was the consequence of very few and small-scale human activities. Slash-and-burn agriculture and the commercial harvesting of timber resulted in the opening of settlements, some of which turned into important cities in the beginning of this century, including Manaus and Belém. Such activities were almost always developed along the Solimoes (as the Amazon is called above its confluence with the Rio Negro) and Amazon Rivers and along the more important tributaries. The rivers were then the natural transport system in the region.

The trend of deforestation changed during the 1960s, with the onset of new human activities induced by the opening of highways that promoted easy and quick access to "terra-firme" areas. The most important of those highways were the Belém-Brasília, the Transamazônica, the Cuiaba-Porto Velho (BR 364) in Rondônia, and the Porto Velho-Manaus-Boa Vista, and more recently (in the 1980s) the railway between Sao Luiz and Carajás. With the expanded transportation system combined with government incentives for development of the region, the population increase was rapid, reaching over 15 million people in the last 2 decades (Sudam, 1987). The main economic activities associated with the occupation process are the following:

o Extensive cattle ranching.
o Timber extraction.
o Perennial crops like cocoa, rubber, and homogeneous forests for pulp and paper.
o Annual crops like sugar cane, soybeans, corn, and rice.
o Charcoal production for the cast iron industry.
o Construction of several major dams for hydroelectric power stations.
o Gold exploration and mining.
o Oil exploration.

For example, extensive cattle ranching has been the major cause of deforestation and has led to many unsuccessful projects both ecologically and economically (Fearnside, 1988). Four industries in the Carajás area are already operating and producing 240,000 tons of cast iron per year, consuming 192,000 tons of charcoal per year. It is estimated that in the next decade production will jump to 280,000 tons per year, consuming 230,000 tons of charcoal per year. With the country's land-tenure criterion, deforested land is a synonym for developed land and real-estate speculation, and even the boldness of the new settler. These processes are all linked to social and economic factors that generate or stimulate migrations, as well as to external factors like the external debt pressure pushing the government to stimulate export policies.

Several of these activities, which lead to deforestation without the introduction of sustainable agriculture or any other real benefit for the population, may be corrected in the near future with the implementation of the program NOSSA NATUREZA (April 1989). This program would regulate the application of fiscal incentives to the development of the Amazon; hopefully, this program will be realized.

THE EXTENT OF DEFORESTATION

One of the most controversial issues recently in the literature, and in the past few months in the Brazilian and international press, has been the extent to which the Amazon region is being deforested. The increasing international concern, as well as the increasing scientific awareness of the role of the tropical forest in future environmental global changes, is the major driving force behind these discussions.

It is now well known that tropical forests are being deforested at increasing rates, but the critical question of knowing with a better accuracy the actual values of such rates is as yet unanswered. Estimates range from predictions that very little primary forest will remain, with the exception of western Amazônia and central Africa, by the turn of the century, to the other extreme that proposes deforestation rates of only 0.6 percent per year (Myers, 1988a). Estimates of deforestation in the Brazilian Amazon range from 5 percent in a recent study by the Instituto Nacional de Pesquisas Espaciais (INPE, 1989) to the 12 percent stated by Mahar (1988), with a more conservative figure of 8 percent given by Fearnside (1989). One of the most striking reports comes from Setzer and Pereira (1989), which shows that over 80,000 km^2 of Amazonian forests were burned in 1987 alone. Comparing this figure with that given by Myers (1988b) of 200,000 km^2 of annual global destruction leaves Brazil as the leading country of tropical deforestation.

When dealing with such controversial numbers, it is important to understand that most of the problems arise from the methodologies now in use to estimate deforestation. Even the use of satellite imagery is liable to errors in interpretation of the data. It is recognized that the National Oceanic and Atmospheric Administration's advanced very high resolution radiometer images with their low resolution (1 km) will have an entire pixel (100 ha) saturated even if only a few hectares are actually burning. Landsat images, although much better as far as resolution is concerned (30 to 100 m), present problems in discriminating primary forest from regrowth and in mapping the borders between forest and savannas. Cloud cover is also a major problem with Landsat images. An assessment using available techniques with greater availability than Landsat images for critical regions like Rondônia and Acre at cheaper prices is urgently needed in order to better evaluate the problem.

Nevertheless, what really counts is knowledge of the rates at which deforestation is occurring, rather than the percentage of deforested areas in the Amazon, whether it be 5 percent or 12 percent. For example, curves fitting the numbers given by Mahar (1988) for several Brazilian states in the Amazônia (Table 15.1) show exponential growth in all cases (Figure 15.1). Critical areas like Rondônia do show an alarming rate, but it is also important to note that areas considered to be noncritical, like Roraima and Amapa, are also showing exponential growth. Assuming the above to be realistic and that the pattern will continue, curve extrapolations show that by the year 2000 most of the Amazonian states will have had their territory completely cleared.

DEFORESTATION AND ITS POSSIBLE PHYSICAL AND BIOLOGICAL CONSEQUENCES

Biodiversity

As stated by Myers (1988a), deforestation "will mean the virtual elimination of what is frequently termed the richest and most complex expression of nature that has ever appeared on the face of the planet. Or, to put it another way, the culmination of almost 4 thousand million years of evolution will have been all but eliminated in less than a century." This statement, besides being alarming, points to the fact that diversity is the basis for the functioning of the tropical forest

TABLE 15.1 Exponential Regression Equations for the Deforestation in Several Amazonian States

State	Total Area (km^2)	Deforested Area, 1988 (km^2)	Equation[a]
Acre	152,589	19,500	$1{,}053 \times 10^{0.0928x}$
Amapa	140,276	572	$117 \times 10^{0.0465x}$
Amazonas	1,567,125	105,790	$413 \times 10^{0.168x}$
Goias	285,793	33,120	$4{,}022 \times 10^{0.0688x}$
Maranho	257,451	50,670	$2{,}758 \times 10^{0.0921x}$
Mato Grosso	881,001	208,000	$1{,}047 \times 10^{0.0969x}$
Para	1,248,042	120,000	$8{,}930 \times 10^{0.0838x}$
Rondônia	243,044	58,000	$1{,}137 \times 10^{0.125x}$
Roraima	230,104	3,270	$41 \times 10^{0.136x}$
Total	5,005,425	598,922	$27{,}600 \times 10^{0.098x}$

NOTE: Data were extracted from Mahar (1988) and cover a range of measurements taken between 1975 and 1988.

[a]The equations are given in the form $y = a \times 10^{bx}$ where x is an integer starting from 1975.

ecosystem. The higher the number of species present in one environment, the higher are the probabilities of adaptation to changes and adversities of the environment.

For the Amazon region, tree diversity has been documented in the literature. Prance et al. (1976) found 505 species higher than 2.5 m in 0.2 ha in a "terra firme" forest near Manaus. Schubart (1982) referred to between 300 to 500 species with breast height diameter (BHD) greater than 5 cm in 1 ha. Martinelli et al. (1988), working in a "terra firme" forest in Rondônia, found about 210 species with BHD greater than 10 cm in 1 ha and estimated a biomass of approximately 400 tons/ha. The latter authors also pointed out the difficulties of estimating biomass for tropical forests. Very little information is available on the chemical composition of the biomass in tropical forests, and this information is urgently needed if the natural biogeochemical cycles are to be understood.

This enormous diversity should serve as the basis for not considering tropical forests as mere sources of wasteful exploitation; much more attention should be paid to what this environment represents to our planet and to the necessity to preserve it. Apart from soil protection and balance of the hydrological cycle, forests are not only sources of timber and fuel wood, but they also can provide many other products like resins, essential and edible oils, fruits and nuts, natural fibers, and

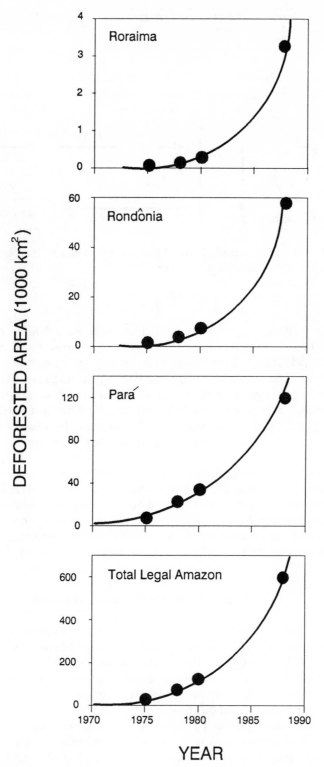

FIGURE 15.1 Deforestation trends in selected areas of the Amazon region. (Data extracted from Mahar, 1988.)

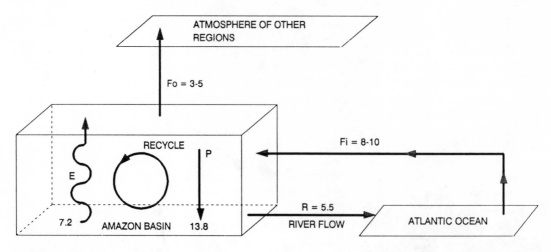

FIGURE 15.2 A schematic model of the water cycle in the Amazon basin.
E is evaporation; P is precipitation.

pharmaceuticals. Many of the drugs that are routinely bought at drug
stores--analgesics, tranquilizers, diuretics, and so forth--are derived
from alkaloids and other biochemicals found in tropical forest plants.
Two drugs derived from the rosy periwinkle, a native of Madagascar's
forests, are being largely and successfully used against Hodgkin's
disease, leukemia, and other blood cancers. The perennial species of
wild corn that is found in a montane forest in Mexico (Iltis et al.,
1979) and is resistant to several viruses and mycoplasms might be the
basis of new commercial cultivars and therefore could expand corn cultivation areas, representing multibillion-dollar savings for mankind. This
was probably the last habitat of the species; it would have been lost
forever with uncontrolled deforestation. That has been the fate of
several endogenous species in places like Rondônia, one of the richest
areas in the world in terms of flora and fauna (Brown, Jr., 1987), where
a substantial part of the region has already been completely cleared.

Hydrological Cycle and Climate

Figure 15.2 schematically shows the water balance for the Amazon
basin, including an area of approximately 6 million km^2 from the river
mouth at Marajo Island to the headwaters at "Cordilheira dos Andes." The
data shown in Figure 15.2, although carrying uncertainties and errors due
to the lack of a measuring network with a good spatial distribution over
the basin, were obtained through successive approaches using different
methods and techniques (Salati, 1986). The most important fact, however,
is to note that the water vapor flux originating in the Atlantic Ocean is
not of sufficient magnitude to explain the rainfall and the vapor outflux
in the basin. As a direct consequence, it is necessary to assume the
recirculation of evapotranspired water in the basin. This conclusion is

strongly supported by the spatial distribution of the ^{18}O and D isotopes in the region (Salati et al., 1979). Other conclusions are that the present atmospheric steady state equilibrium in the region is dependent on the vegetation cover, i.e., the forest. Further, the Amazon basin is a source of water vapor to other regions, especially to the Brazilian central plateau, and eventually to the "Pantanal" (a large area of swamp and seasonally flooded land in central-western Brazil).

Bearing such evidence in mind, it is clear that a major alteration in vegetation cover of the region would lead to changes in the climate at the micro- and meso-levels. Changes would be felt particularly through variations in the albedo, in the rainwater residence time, and in an increase in runoff and a decrease in evapotranspiration (Salati et al., 1979). Increases in maximum temperatures and daily thermal amplitudes, due to a decrease in precipitation, would also be expected. Such alterations would be felt not only in the Amazon region itself but also in other nearby regions, especially the Brazilian central plateau. Simulation models, although showing several uncertainties, also point to such a scenario (Henderson-Sellers, 1987).

Transport of Sediments

Soil erosion is one of the most commonly cited consequences of deforestation, particularly in tropical areas where precipitation is high. Besides being directly responsible for a reduction in soil fertility, erosion also plays a major role in the decrease of the lifespan of hydroelectrical power dams. For example, although it was difficult to predict at the time, the sedimentological studies done for the construction of the Samuel power dam in Rondônia did not take into account the possible changes in land use and their consequences for the sediment load of the Rio Jamari. Studies that are currently being carried out using the ^{210}Pb technique to calculate the sedimentation rate in a lake near the dam are showing increases in the rates that might be closely correlated with deforestation and/or tin mining activities in the basin (B. R. Forsberg, personal communication). These data are, however, very preliminary and urgently need further confirmation.

Although consequences as just exemplified could actually be foreseeable, their prediction is impaired by lack of data. There are very few studies of land erosion and river sediment loads in tropical areas. The little existing data do show, however, that erosion losses can be 100 times greater in soils changed to agricultural use when compared to similar soil covered with forest (Salati and Vose, 1984). This lack of data is in reality a result of the difficulties in applying conventional methods of erosion measurement to tropical areas. Alternatively, the sediment flux per unit area of a river basin, the so-called denudation rate, can be used to estimate erosion losses. There are, however, doubts about the accuracy of such methods. Some authors argue that several depositional areas may exist in a basin, which would lead to long lag times in the response from the river to changes in land use (Trimble, 1975; Meade, 1989). Graham, Jr. (1986), for instance, working in the Rio Jamari, estimated erosion losses by using both the classical Universal

Soil Loss Equation (USLE) and measurements of the sediment load in the river from 1978 to 1983. Both methods showed an increase in erosion, the latter, however, giving smaller numbers than the first. The ratio between estimates calculated by the denudation rate and the ones calculated through the USLE was on average 0.06 and was highly variable, ranging from 0.01 to 0.12. Although differences observed between the methods were great, it is important to note that both presented similar trends and provide a basis for long-term estimates.

Since 1982, as a result of a joint cooperation program between the Instituto Nacional de Pesquisas na Amazônia (INPA), the Centro de Energia Nuclear na Agricultura (CENA), the Escola Superior de Agricultura Luiz de Queiroz (ESALQ), and the University of Washington, sediment concentrations and fluxes have been systematically measured in the Amazon/Solimoes main channel and several of the major tributaries, including the Rio Madeira and some of its tributaries. Table 15.2 summarizes the available data. Details about sample collection, analysis, and other specific information may be found in Meade et al. (1985), Meade (1985), Richey et al. (1986), Mertes (1985), and Martinelli et al. (1989).

These data are, however, preliminary for use in deforestation monitoring; greater knowledge of their spatial and temporal variability is needed. The results do show the general expected trend for denudation rates in the basin. Rivers with headwaters in the Andean and sub-Andean regions have the highest values, and tributaries originating in the Brazilian Plateau show the lowest values. It should be noted, however, that Rondônia rivers carry very little sediment and could therefore be useful if used as monitors of the rapid deforestation that is occurring in their basins.

Carbon Dioxide Emissions

Changing the forest cover to other land use systems transfers substantial quantities of carbon stored in the biomass to the atmosphere, potentially increasing the greenhouse effect (Houghton et al., 1987). Since 1988, when the massive forest burnings in Rondônia generated enormous pressure by the international media, many scientists started worrying about obtaining better estimates of the contribution of forest burnings to CO_2 emissions to the atmosphere.

Here we calculate Amazônia's potential contribution to atmospheric CO_2. We made several simplifying assumptions: (1) biomass will always be completely burned, (2) successional ecosystems will always have a small biomass when compared to the original forest (this is true taking into consideration that it takes about 100 years to completely regenerate a forest to its original status in the region), and (3) soil organic matter will not increase significantly with time after burning. Variability in the calculations arises from the uncertainty in biomass estimation for the Amazon forest. We used the values found by Martinelli et al. (1988) of 360 ± 60 tons/ha of above-ground biomass plus 40 tons/ha of litter and fallen trunks for the environmental protection area of the Samuel power dam in Rondônia. Total biomass available for burning could therefore vary between 280 and 400 tons/ha, which would give from 140 to

TABLE 15.2 Summary of Available Data on Water Discharge, Sediment Flux, and Denudation Rate of the Amazon and Madeira River Basins Between 1982 and 1986

River Basin	Discharge (10^3 m^3/s)	Total Suspended Sediment (10^6 tons/year)	Denudation Rate (tons/km^2 per yr)
Amazon			
Vargem Grande	47.9	647	571
Obidos	159.0	1156	250
Tributaries			
Içá	7.4	19	127
Jutaí	1.4	2	23
Juruá	3.0	26	116
Japurá	13.9	23	79
Purús	10.8	29	78
Negro	30.8	6	8
Madeira			
Main channel	29.2	488	357
Tributaries			
Pimenta Bueno	0.245	0.17	14
Jiparaná	0.875	0.75	23
Jaru	0.081	0.03	6

200 tons of carbon per hectare if it is assumed that biomass contains 50 percent carbon. The final numbers used in our calculations are the total deforested area and the annual rate of deforestation. These numbers are highly controversial, as already stated, and the total deforested area ranges from the very optimistic estimate of 250,000 km^2 (5 percent of the Legal Amazon--INPE, 1989) to the other extreme of 600,000 km^2 (12 percent of the Legal Amazon--Mahar, 1988). As far as annual deforestation rates are concerned, the numbers given for 1988 range from 17,000 km^2/yr (INPE, 1989) to 80,000 km^2/yr (Setzer and Pereira, 1989).

Taking the above cited numbers and assumptions into account, we estimated that the Amazon region has already contributed emissions ranging from 3.5 X 10^{15} g carbon to 12 X 10^{15} g carbon to the atmosphere (Table 15.3). That is on the order of 2 to 7 percent of the total atmospheric CO$_2$ emitted due to deforestation and burning until 1980 (Woodwell, 1987). Table 15.3 also shows that annual emissions, considering the 1988 rates, ranged from 0.24 X 10^{15} to 1.6 X 10^{15} g carbon per year, which is between 4 to 25 percent of the global CO$_2$ emissions, estimated to be 7 X 10^{15} g carbon per year (Woodwell, 1987). Even if, to be on the conservative

TABLE 15.3 Estimated CO_2 Emission due to Burning in the Amazon

Estimate of Carbon Biomass Available[a] (tons/ha = 10^8 g/km^2)	Range of CO_2 Emissions (X 10^{15} g carbon)	
	Cumulative Total[b]	Total in 1988[c]
Lower (140)	3.5 to 8.4	0.24 to 1.1
Upper (200)	5.0 to 12.0	0.34 to 1.6

NOTE: Estimate based on the assumption that 100 percent of the burned biomass is transformed into CO_2.

[a]Based on data from Martinelli et al. (1988).

[b]The total range of emissions is calculated as the product of the lower and upper estimates of the carbon biomass available and the lower (250,000 km^2--INPE, 1989) and upper (600,000 km^2--Mahar, 1988) estimates of the total area deforested.

[c]The range of emissions for 1988 is calculated as the product of the lower and upper estimates of the carbon biomass available and the lower (17,000 km^2--INPE, 1989) and upper (80,000 km^2--Setzer and Pereira, 1989) estimates of the area deforested in 1988.

side, we assumed that only 50 percent of the biomass is burned during forest fires, the numbers would still be significant, ranging from 2.0 to 12 percent of global emissions. Although these data may not be precise, they are useful as a warning for the potential dangers that uncontrolled deforestation in the Amazônia poses to our environment.

CONCLUSIONS AND RECOMMENDATIONS

Present knowledge reveals the key role of the forest in maintaining the dynamic equilibrium of the Amazonian ecosystem. In summary, the forest controls water dynamics in the basin, the energy balance, the sediment yield, the nutrient balance, the diversity of species, the quality of surface water, the soil quality, and the carbon stock in the biosphere.

The above conclusions were drawn from an analysis of many papers published about the Amazônia in the last few years. Although the information available is relevant and permitted those conclusions, many doubts still persist, because with rare exceptions the information is not continuous in time. Spatial distribution of the available information is also poor. It is therefore still difficult to visualize the entire Amazonian ecosystem as a whole. The lack of knowledge of the basic functioning mechanisms of the Amazonian ecosystem and of the most suitable methods for sustainable development of the region are the main reasons for the failure of many of the agricultural and cattle ranching projects. In addition, institutional problems are serious. The lack of research funds is impairing the continuity of many research programs and also the implementation of technical recommendations and enforcement of legal exigencies. We therefore recommend the following:

o An increase in the number of biological conservation areas in regions representative of different ecosystems, in order to preserve endogenous species from extinction.

o An expansion of international support to research groups and institutions dedicated to the study of the basic functioning mechanisms of the ecosystem, especially those with programs already under way for a considerable time.

o The design of integrated interdisciplinary regional programs to study the factors affecting the steady state equilibrium of the entire ecosystem.

o The design of programs to reclaim already degraded areas, particularly through reforestation.

o The implementation of anthropology research programs designed to acquire better knowledge of the forest management practices of native peoples. We should not forget that they have been living in the forest and from the forest for thousands of years.

o The implementation of programs for the protection of the native Indian communities and their culture and traditions.

o The stimulation of extractive activities for natural products, apart from logging, that have already proved to be economically profitable and that cause no harm to forest integrity.

ACKNOWLEDGMENTS

The authors wish to acknowledge the financial support to Eneas Salati by the Smithsonian Institution during the preparation of this paper. This paper is contribution No. 35 from the National Science Foundation CAMREX project and contribution No. 6 of the International Atomic Energy Agency Amazônia I project (IAEA-BRA/0/010).

REFERENCES

Brown, Jr., K.S. 1987. Soils and vegetation. In: Whitmore, T.C. and Prance, G.T. (eds.) Biogeography and Quaternary History in Tropical America. Clarendon Press, Oxford.

Fearnside, P.M. 1988. Causas do desmatamento na Amazônia Brasileira. Para Desenvolvimento, Meio Ambiente 23:24-33.

Fearnside, P.M. 1989. Deforestation in Brazilian Amazônia. In: G.M. Woodwell (ed.) The Earth in Transition: Pattern and Processes. Biotic Impoverishment. Cambridge University Press, New York (in press).

Graham, Jr., D.H. 1986. The Samuel Dam: Land Use, Soil Erosion, and Sedimentation in Amazônia. Master's Thesis, University of Florida.

170

Henderson-Sellers, A. 1987. Effects of change in land use on climate in the humid tropics. In: Dickinson, R.E. (ed.) The Geophysiology of Amazônia. John Wiley & Sons, New York.

Houghton, R.A., Boone, R.D., Frucci, J.R., Hobbie, J.M., Melillo, J.M., Palm, C.A., Peterson, B.J., Shaver, G.R., and Woodwell, G.M. 1987. The flux of carbon from terrestrial ecosystems to the atmosphere in 1980 due to changes in land use: geographic distribution of the global flux. Tellus 39:122-139.

Houghton, R.A., Boone, R.D., Melillo, J.M., Palm, C.A., Woodwell, G.M., Myers, N., Moore, B., and Skole, D.L. 1985. Net flux of carbon dioxide from tropical forests in 1980. Nature 316:617-620.

Iltis, H.H., Doebley, J.F., Guzman, R.M., and Pazy, B. 1979. Zea diphloperennis (Gramineae), a new Teosinte from Mexico. Science 203:186-188.

Instituto de Pesquisas Espaciais (INPE). 1989. Avaliaçao da alteraçao da cobertura florestal na Amazônia utilizando sensoriamento remoto orbital. Primeria ediçao, Sao José dos Campos.

Martinelli, L.A., Brown, I.F., Ferreira, C.A.S, Thomas, W.W., Victoria, R.L., e Moreira, M.Z. 1988. Implantaçâo de Parcelas para Monitoramento de Dinâmica Florestal na Area de Proteçâo Ambiental, UHE Samuel, Rondônia. Relatório Preliminar. 1988.

Martinelli, L.A., Forsberg, B.R., Victoria, R.L., Devol, A.H., Mortatti, J., Ferreira, J.R., Bonassi, J.A., and de Oliveira, E. 1989. Suspended sediment load in the Madeira River. In: Degens, E.T., Kempe, S., and Eisma, D. (eds.) Transport of Carbon and Minerals in Major World Rivers, Lakes and Estuaries, Pt. 6. Mitt. Geol.-Palaont. Inst., Univ. Hamburg, SCOPE/UNEP Sonderband 68 (in press).

Mahar, D.J. 1988. Government Policies and Deforestation in Brazil's Amazon Region. A World Bank Publication in cooperation with the World Wildlife Fund and The Conservation Foundation. Washington, D.C.

Meade, R.H. 1985. Suspended Sediments in the Amazon River and Its Tributaries in Brazil During 1982-1984. U.S. Geological Survey Open File Report 85-492.

Meade, R.H. 1989. Movement and storage of sediment in river systems. In: Lerman, A., and Meybeck, M. (eds.) Physical and Chemical Weathering in Geochemical Cycles. Reidel Press, Dordrecht, The Netherlands (in press).

Meade, R.H., Dunne, T., Richey, J.E., Santos, U.M., and Salati, E. 1985. Storage and remobilization of suspended sediment in the lower Amazon River of Brazil. Science 228:488-490.

Mertes, L.A.K. 1985. Floodplain Development and Sediment Transport in the Solimoes-Amazon River, Brazil. M.S. Thesis, University of Washington.

Myers, N. 1988a. Natural Resources Systems and Human Exploitation Systems: Physiobiotic and Ecological Linkages. The World Bank Policy Planning and Research Staff, Environment Department, Environment Department Paper No. 12.

Myers N. 1988b. Tropical deforestation and remote sensing. Forest Ecology and Management 23: 215-225.

Prance, G.T. 1986. The Amazon: Paradise Lost? In: Kaufman, L., and Mallory, K. (eds.) The Last Extinction. MIT Press, Cambridge, Mass.

Prance, G.T., Rodrigues, W.A., and da Silva, M.F. 1976. Inventario florestal de um hectare de mata de terra-firme Km 30 Estrada Manaus-Itacoatiara. Acta Amazonica 6:9-35.

Richey, J.E., Meade, R.H., Salati, E., Devol, A.H., Nordin C.F., and dos Santos, U.M. 1986. Water discharge and suspended sediment concentration in the Amazon River. Water Resources Research 22:756-764.

Salati, E. 1986. The Climatology and Hydrology of Amazônia. In: Prance, G.T., and Lovejoy, T.E. (eds.) Amazônia. Pergamon Press, Oxford.

Salati, E., and Marques, J. 1984. Climatology of the Amazon region. In: Sioli, H. (ed.) The Amazon Limnology and Landscape Ecology of a Mighty Tropical River and Its Basin. Dr. W. Junk Publishers, Dordrecht, The Netherlands, pp. 87-126.

Salati, E., and Vose, P.B. 1984. Amazon basin: A system in equilibrium. Science 225:138-144.

Salati, E., Dall Ollio, A., Gat, J., and Matsui, E. 1979. Recycling of water in the Amazon basin: an isotope study. Water Resources Research 15: 1250-1258.

Setzer, A.W., and Pereira, M.C. 1989. Amazon biomass burning in 1987 and their atmospheric emissions. Science (in press).

Schubart, H.O.R. 1982. Fundamentos ecológicos para o manejo florestal na Amazônia. Silvicultura em Sâo Paulo 16A:713-731.

Superintendencia do Desenvolvimento da Amazônia (Sudam). 1987. Censos Demográficos das Unidades que compôem a Amazônia Legal. Relatorio da Divisáo de Estatística da Sudam, Brasilia.

Trimble, S.W. 1975. Denudation studies: Can we assume stream steady state? Science 188:1207-1208.

Woodwell, G.M. 1987. The warming of the industrialized middle latitudes 1985-2050: Causes and consequences. Developing Policies for Responding to Future Climatic Changes. Villach, Austria, September.

Woodwell, G.M., Wittaker, R.H., Reiners, W.A., Likens, G.E., Delwiche, C.C., and Botkin, D.B. 1978. The biota and the world carbon budget. Science 199:141-146.

16

IMPACTS OF GLOBAL CHANGE

Jose Goldemberg

In reality global change in some areas is both a cause and a consequence, particularly in agriculture and industry: 14 percent of the greenhouse gases originate in agriculture, 3 percent in industry (plus the 17 percent due to chlorofluorocarbons (CFCs)), and 9 percent in modifications of land use and deforestation. The remaining 57 percent originate in the industry of energy production.

A discussion of impacts is important to determine priorities for action: How real and important impacts are will make us move more rapidly or slowly. This was clearly the case with CFCs. As soon as the dramatic impacts on the ozone layer in the Antarctic were clearly demonstrated, governments and the diplomatic establishment--often described as slow and inefficient--moved quickly and agreed on the Montreal Protocol.

The dilemma of most people facing impacts of any kind is whether to take preventive measures to eliminate them or adaptive and corrective measures to live with them. What one does depends frequently on relative costs, on who pays for changes, and on the time frame of the changes. Frequently one acts only when it is too late or too expensive to take preventive measures.

A good example is the cost of afforestation as compared to the cost of deforestation and the advantages one gets from deforestation. Typically it costs U.S.$1000 to reforest 1 hectare, and it is very dubious that one gets more than that by destroying the forest for short-term gain. In the Amazon forest, a good part of the 25 million hectares burned so far (at a rate of approximately 2 million hectares per year), has proven to be unfit for sustained agriculture. This seems to be a very unproductive way to go. It is therefore in the self-interest of Brazilians to prevent that from happening at the risk of contributing some 7 percent to the amount of carbon thrown into the atmosphere per year.

The reason we are so keen on discussing global change these days is that we want to prevent it before it is too late.

In dealing with such problems, externalities are what really count, and one cannot rely on market forces. Government intervention is accepted naturally, and we see more and more of it happening all over the world. The recent example of Los Angeles, where the local Environmental Board decided to clean up the city at a cost of some $3 billion per year (U.S.$250 per person per year), is very impressive.

172

Our responsibility therefore is to characterize clearly and convincingly the problems or impacts. When the problems are clear cut, governments move quickly.

Some impacts are not very clear, and the usual overcautious position adopted by scientists is frequently used by governments as an excuse for not acting. One example of an impact about which we could be more aggressive than we are today is energy production. Fifty-seven percent of the greenhouse effect is due to carbon dioxide and other gases emitted in the burning of fossil fuels. One could, in principle, face the impacts of this part of the greenhouse problem by spending a lot of money to recapture the carbon dioxide in stacks, practice afforestation, or simply cope with higher temperatures and rising sea levels. There are estimates that this approach could be taken with a few hundred billion dollars. This is called adaptive behavior.

The other approach is preventive: one reduces energy production and thereby reduces carbon dioxide emissions. It has been exhaustively demonstrated that this is the cheaper road to take and that it can be done in industrialized countries through improved efficiency in the production and use of energy.

It is not clear, however, that the same can be done in developing countries, where energy is so essential for development and where economic activities are bound to grow much more than in the industrialized nations. Estimates indicate that by the year 2020, two-thirds of the energy consumed in the world will be consumed in the less-developed countries, up from the one-third presently being used.

Therefore the magnificent opportunity lies ahead to steer this evolution in the right direction, introducing in the developing countries energy systems that have built in the improved energy technologies already available in the industrialized countries. In doing so, the less-developed countries could leapfrog the painful adaptation going on in the developed part of the world and make an important contribution to reducing the impacts of energy use on the biosphere.

PART D   IMPLICATIONS FOR PUBLIC POLICY

17

THE GLOBAL ENVIRONMENT: A NATIONAL SECURITY ISSUE*

Albert Gore, Jr.

Many have come to share the belief that humankind has suddenly entered into a brand-new relationship with the planet Earth and that human civilization is, in its current pattern, causing grave and, perhaps soon, irreparable damage to the ecological system that supports life as we know it. My purpose is to sound an alarm--loudly and clearly--of imminent and grave danger, and to describe a strategy for confronting this crisis, with changes in our collective behavior and thinking that, if made, can forestall and prevent the horrendous prospect of an ecological collapse.

First, why is such an alarm necessary? Do we need a crisis before we can act? Sometimes in human affairs a pattern is well set before its implications are felt in our daily lives. This is true both in politics and in science. When shattered glass filled the streets of Berlin on Kristallnacht, few could conceive of the holocaust to follow. But from a distance, the pattern is now clear. When the first atom was split, few could conceive of nuclear bombs. But when Albert Einstein wrote to Franklin Roosevelt, the pattern was clear. How much information is needed by the human mind to recognize a pattern? How much more is needed by the body politic to justify action in response? It took a long time for the world to respond to Adolf Hitler. Because of Hitler, it took only a short time for Roosevelt to respond to Einstein.

In a classic experiment often cited, a frog dropped into a pot of boiling water quickly jumps out. But the same frog, put into the water before it is slowly heated, will remain in place until it is boiled. The meaning of a pattern is conveyed by contrast as opposed to sameness. Sameness lulls the senses and conveys an absence of danger. Gradual change sometimes resembles sameness, obscuring danger from minds that reserve their alertness for sharp contrasts. Exponential change at first resembles sameness, then gradual change, then explosive contrast. It is often hard to recognize the shape of an exponential curve before it reaches the explosive stage. It is difficult because the contrast essential to understanding very large patterns is sometimes visible only from a distance.

---

*Presented May 1, 1989, as a speech to the assembled speakers for the Forum on Global Change and Our Common Future.

If an individual or a nation is accustomed to looking at the future 1 year at a time and the past in terms of a single lifetime, then many large patterns are concealed. If a political body looks at policies in the context of a single nation, then the global impacts will remain invisible.

In the relationship of the human species to the planet Earth, not much change is visible in a single year, in a single nation. Yet if one looks at the entire pattern of that relationship from the emergence of the species until the present, a distinctive contrast in very recent times clearly conveys the danger to which we must respond. It took 10,000 human lifetimes for the population to reach 2 billion. Now in the course of a single human lifetime, it is rocketing from 2 billion toward 10 billion, and is already halfway there. Startling graphs showing the loss of forest land, topsoil, stratospheric ozone, and species all follow the same pattern of sudden, unprecedented acceleration in the latter half of the twentieth century. And yet, so far, the pattern of our politics remains remarkably unchanged.

The earth's forests are being destroyed at the rate of one football field's worth every second, one Tennessee's worth every year. An enormous hole is opening in the ozone layer, reducing the earth's ability to protect life from deadly ultraviolet radiation. Living species are dying at such an unprecedented rate that more than half may disappear within our lifetime. Chemical wastes, in growing volumes, seep downward to poison ground water and upward to destroy the atmosphere's delicate balance. Huge quantities of carbon dioxide, methane, and chlorofluoro-carbons (CFCs) dumped into the atmosphere are trapping heat and raising global temperatures.

In 1987 carbon dioxide levels in the atmosphere began to surge with record annual increases. Global temperatures are also climbing: 1987 was the second hottest year on record; 1988 was the hottest. Scientists now predict our current course may raise world temperatures almost 5°C in the lifetime of our children. The last time there was such a shift, it was 5°C colder: New York City was under 1 km of ice. If 5°C colder over thousands of years produces an ice age, what could 5°C warmer produce in one lifetime?

Why are these dramatic changes taking place? Because the human population is surging, because the industrial, scientific, and technological revolutions magnify the environmental impact of these increases, and because we tolerate self-destructive behavior and environmental vandalism on a global scale.

The problem in organizing our response is that the worst effects seem far off in the future, and they are so unprecedented they seem to defy common sense, while right now, in the present, millions of people are suffering in poverty and dying because of starvation, warfare, and preventable diseases. How do we deal with these immediate problems and at the same time confront the problems of the future? One of the philosophers of the environmental movement, Ivan Illich, in a recent interview explained the sudden environmental activism of Margaret Thatcher, Mikhail Gorbachev, and other world leaders previously uninterested in the global environment by saying, "What has changed is that our common sense has begun searching for a language to speak about the shadow our future throws."

Science already has such a language. Consider a picture showing how time and space are shaped by mass, with a black hole pictured as a deep well in a grid, with the space and time around the well sloping toward it. Human political awareness is shaped by history in precisely that way. Our political awareness of the world is shaped and bent by events. Large events like World War II exert a powerful gravitational pull on every idea we have about the world around us. The Holocaust shapes every idea we have about human nature. And just as in Einstein's theory, future events can exert the same gravitational pull on our thinking as events in the past--even though the events in the future have not yet occurred.

Time is relative in politics as in science. The political will that made possible the mass political protests against escalating the nuclear arms race came from awareness of a downslope toward a future we did not want to see. Many felt us being pulled toward a nuclear war that would crush human history forever into a black hole. We are now changing our course away from that downslope, we hope, and taking a new direction-- even though 99.99 percent of all human beings on earth have never seen, heard, or personally felt nuclear destruction. The awareness of that potential future event came from political communication, with abstract symbols, like words.

Now throughout the world we are witnessing the emergence of a new political will to take a different course in order to avoid the slope toward global environmental destruction. We see the catastrophe coming; we hear Rachel Carson's silent spring. The slope seemed gradual at first but now it is steep. We feel strongly pulled toward ecological collapse by the policies we are now pursuing.

I personally became deeply involved in the effort to avoid a nuclear holocaust 9 years ago because I felt the slope toward that horrendous possibility. And I tried to bring to the task the skills of my pro- fession. I believe all the talk about the global environment as a national security issue makes a great deal of sense in political terms.

For the past 13 years, as a citizen and as a member of the U.S. Congress, I have had long-standing interests in both the environmental threat and in national security. As a practical matter I have dealt with these subjects as separate intellectual accounts involving distinct areas of public policy, each with its own completely different set of concerns and participants. Yet they grow more and more alike. National security comprises matters that directly and imminently menace the interests of the state or the welfare of the people. As such, these issues command the attention of political leaders at the highest level, with a propor- tionate claim on the resources of government and the wealth of the nation. If society were an organism, national security would involve the instinct for survival.

To this point, the national security agenda has been dominated by issues of military security, embedded in the context of global struggle between the United States and the Soviet Union, a struggle that the protagonists have often waged through distant surrogates but that has always harbored the risk of direct confrontation and nuclear war. Given

the changes in Soviet behavior that have begun under Gorbachev, there is
growing optimism that this long, dark period may be passing. There is
also hope that this will open the international agenda for other urgent
matters and for the release of enormous resources, now committed to war,
toward other objectives.

Many hope that the global environment will be the new dominant issue.
They assert that a collective, international struggle for stability in
the ecosystem will succeed the old pattern of national struggle for temporal power and will justify the preemption of enormous resources, and
reshape the public consciousness in support of another long, global
struggle.

I am deeply in sympathy with this view, and yet as someone who has
worked hard on both issues, I believe the analogy must be used very
cautiously. The U.S.-Soviet struggle has lasted almost half a century,
consumed several trillion dollars, cost close to 100,000 American lives
in Korea and Vietnam, and profoundly shaped the psychological and social
consciousness of our people. Much the same could be said of the Soviets,
who, if anything, have endured far more than have we for the sake of
their ideology.

Nothing is automatic or foreordained about the course of U.S.-Soviet
relations, no matter how many editorial writers now claim, "The Cold War
is over." Nothing relieves us of our present responsibilities for defense or of the need to conduct painstaking negotiations to limit arms
and reduce the risk of war. The old agenda is with us still, exacting
its price in wealth, creativity, and the attention of statesmen.

And yet, the environmentalists are right.

Certainly, there is strong evidence that the new enemy is at least as
real as the old. For the general public, the shocking images of last
year's drought, or of beaches covered with medical garbage, inspired a
sense of peril once sparked only by Soviet behavior. And for environmental specialists, the steady flow of data from scientific investiga-
tion of the environment--often ambiguous, but always menacing--is eerily
equivalent to intelligence collection against the more familiar Soviet
threat. The U2 spy plane, for example, now is used to monitor not
missile silos, but ozone depletion.

Already we are seeing governments struggling to resolve issues whose
domains go far beyond anything in our experience. Debate over the disposition of radioactive wastes, for example, involves choices that must
remain valid across geological time. The species now disappearing at an
unprecedented rate will never return. The global climate pattern could
shift to a new equilibrium and never regain its former pattern.

In the not distant future, there will be a new "sacred agenda" in
international affairs, policies that enable the rescue of the global
environment. This task will one day join, and then perhaps even sup-
plant, preventing the world's incineration through nuclear war, as the
principal test of statecraft.

However, in thinking about environmentalism as a national security
concern, it is important to differentiate between what would--in military
jargon--be called the level of threat. Certain environmental problems
may be important but are essentially local, others cross borders and in
effect represent theaters of operations, and still others are global and

strategic in nature. On this scale, even phenomena as important as the slow suffocation of Mexico City, the deaths of northern forests in America and Europe, or even the desertification of large areas of Africa will likely not be regarded as full-scale national security issues.

However, the greenhouse effect and stratospheric ozone depletion fit the profile of national security issues of global significance. These phenomena certainly will in time produce effects big enough to threaten international order, even at the level of war and peace. In the case of global warming, the fact that some of the worst effects will not fully manifest themselves until the middle of the next century is offset by the fact that actions we take now will determine the extent of the damage later.

When nations perceive that they are threatened at the strategic level, they may be induced to think of drastic responses involving sharp discontinuities from everyday approaches to policy. In military terms, this is the point when the United States begins to think of invoking nuclear weapons. The global environment may well involve responses that are, in comparative terms, just as radical--not just business as usual, not just incremental variations, but massive departures from the norm.

Nuclear war is an apocalyptic subject, and so is global environmental destruction. We are dealing here with increasingly credible forecasts of climatic dislocations, vast changes in growing cycles, inundations of coastal areas, and the loss to the sea of vast territories--some of them very heavily developed and populated. We are dealing not only with a threat to human health, but also with unpredictable and potentially vast changes to all life at the surface of the earth and the seas as the result of prolonged exposure to increased ultraviolet radiation.

What is more, despite some progress made toward limiting some sources of the problem, such as CFCs, we have to face the stark fact that we have barely scratched the surface. Even if all other elements of the problem are solved, a major threat is still posed by emissions of carbon dioxide, the exhaling breath of the industrial culture upon which our civilization rests. The implications of the latest and best studies on this matter are staggering. We must be honest about them. Essentially, they tell us that with our current pattern of technology and production, we face a Hobbesian choice between economic growth in the near term and massive environmental disorder as the subsequent penalty.

This central fact cuts across the face of all environmental strategies as we generally think of them. It suggests that the notion of environmentally sustainable development at present may be an oxymoron rather than a realistic objective. It declares war, in effect, on routine life in the advanced industrial societies. And--central to the outcome of the entire struggle to restore global environmental balance--it declares war on the Third World.

The Third World does not have a choice about whether or not it will develop economically. If it does not develop economically, poverty, hunger, and disease will consume entire populations. And long before that, whole societies will experience revolutionary political disorder. Rapid economic growth is a life-or-death imperative throughout the Third World. The peoples and governments of the Third World will not be denied that hope, no matter what the longer-term costs are for the global environment.

And why should they accept what we, manifestly, will not accept for ourselves? Who is so bold as to say that any nation in the developed world is prepared to abandon industrial and economic growth? Who will proclaim that any nation in the developed world will accept even serious compromises in levels of comfort for the sake of global environmental balance? Who will apportion these sacrifices? Who will then bear them? Development, of course, is part of the problem as well as the solution. We know that, just as we know that nuclear deterrence depends on the weapons we are trying to render obsolete.

The effort to solve the nuclear arms race has been complicated not only by simplistic stereotypes of the enemy and the threat he poses. It also has been complicated by simplistic demands for immediate unilateral disarmament, without any basis for a widely shared confidence that the original threat is no longer real.

My own belief is that perceptions must evolve simultaneously in both superpowers, keeping pace with changing technology, accompanied by conscious efforts to improve information each side has about the other and about the nature of the threat--all aimed at increasing mutual confidence that the threat is in fact changing and receding. In similar fashion, the effort to solve the global environmental crisis will be complicated not only by blind assertions that more and more environmental manipulation and more and more resource extraction are essential for economic growth. It will also be complicated by simplistic demands that development, or technology itself, must be stopped for the problem to be solved. This is a crisis of confidence that must be addressed.

Recently, when our son was hit by an automobile, my wife and I lived in the world of medical science, pursuing the goal of restoring our son to health. Ten years ago, a child with his injuries would have had several surgical interventions that doctors now realize are unwise, given the greater likelihood of the body healing itself where some injuries are concerned. Yet two days after the accident, when doctors were unable to stabilize internal bleeding, my wife and I naively hoped that some natural healing process would take care of that problem and make any surgery unnecessary. If the doctor had relied not on science but on us, we would have unwisely urged against an operation. But doctors have acquired sufficient knowledge to realize that the body should be allowed to heal itself when it can do so. At the same time, they know there are some instances in which intervention is essential to save life.

Similarly, we must acquire sufficient knowledge of the earth's system to judge when it can heal itself and when it is necessary for us to intervene. For example, when 40,000 children die of disease and starvation every 24 hours, we obviously must intervene. But it is past time to recognize that many of society's interventions in the environment have been and are unwise. Much ecological destruction is subsidized by governments. We need more knowledge, more experience, and the kind of sensitive judgments that modern doctors have learned to make.

The cross-cut between the imperatives of growth and the imperatives of environmental management represents a supreme test for modern industrial civilization. The test is whether we can devise very dynamic new strategies that will accommodate economic growth within a stabilized environmental framework. That is an extreme demand to place on technology. There is no real assurance that such a balance can in fact be struck. Nevertheless the effort must be made. And because of the urgency, scope, and even the improbability of complete success in such an endeavor, I am strongly tempted to use a military term for the metaphor. To deal with the global environment, we will require the environmental equivalent of the Strategic Defense Initiative (SDI), a "Strategic Environment Initiative."

I have been an opponent of the military SDI. But even opponents of the SDI recognize that this effort has been remarkably successful in drawing together previously disconnected government programs, in stimulating the development of new technologies, and in forcing upon us a wave of intense new analysis of subjects previously thought to have been exhausted.

We need the same kind of focus and intensity, and similar levels of funding, to deal comprehensively with global warming, stratospheric ozone depletion, species loss, deforestation, ocean pollution, acid rain, air and water pollution, and all of the problems degrading the world's environment. In every major sector of economic activity--energy, agriculture, manufacturing, and transportation, for example--a Strategic Environment Initiative (SEI) must identify and then spread sets of increasingly effective new technologies--some that are already well in hand; some that need further work, although they are well understood in principle; and some that are revolutionary ideas whose very existence is now a matter of speculation.

Let me briefly illustrate. Energy is the lifeblood of development. Unfortunately, today's most economical technologies for converting energy resources into usable forms of power--as in burning coal to make electricity--release a plethora of pollutants. An "Energy SEI" should focus on producing energy for development without compromising the environment. Chief on the near-term list of alternatives are energy efficiency and conservation; on the mid-term list, solar power, possibly new-generation nuclear power, and biomass, as well as enhanced efficiency; and on the long-term list, nuclear fusion, as well as enhanced versions of solar, biomass, and nuclear energy, and energy efficiency.

In agriculture, we have witnessed vast growth in Third World food production through the Green Revolution, but often that growth relied on heavily subsidized fertilizers, pesticides, irrigation, and overall mechanization, sometimes giving the advantage to rich farmers over poor ones. We need a second green revolution to address the needs of the Third World's poor. An "Agricultural SEI" must focus on increasing productivity from small farms on marginal land, and on further development of low-input agricultural methods. These advances, whose components are not only technological, but financial and political as well, may be the key to satisfying the land hunger of the disadvantaged and the desperate who are slashing daily into the rain forest of Amazonia--leaving behind the depleted soil of their previous homesteads. They may also be the key in the battle to arrest the desertification of sub-Saharan Africa, where human need and climate stress are now operating in a deadly partnership.

184

Fortunately, the next wave of agricultural improvements is almost upon us--from biotechnology. In my view, we should carefully push forward work on new crop strains with genetic encoding that allows "natural" resistance to pests, disease, and droughts, not to mention improved yield. Of course, biotechnology will not completely solve the problems that arise from inadequate distribution of food supplies--they are most often due to a failure of politics, not crops.

In addition, new industrial processes, new materials, and increased use of recycled materials will all become important to "sustainable" development.

Needed in the United States probably more than anywhere is a "Transportation SEI" focusing in the near term on improving the mileage standards of our vehicles and encouraging and enabling Americans to drive less. In the mid-term come questions of alternative fuels, such as biomass-based liquid fuels or electricity. And in the mid-term and long term comes the inescapable need for reexamining the entire structure of our transportation sector and its inherent demand on the personal vehicle for efficient transport.

Funds to promote these research objectives could be drawn from very modest U.S. energy taxes, eventually perhaps even a carbon dioxide (CO_2) tax, although that is not yet politically viable. The U.S. government should organize itself to finance the export of energy-efficient systems and renewable energy sources. That means preferential lending arrangements through the Export-Import Bank and Overseas Private Investment Corporation.

Encouragement for the Third World should also come in the form of attractive international credit arrangements for energy-efficient and environmentally sustainable processes. Funds for this lending stream would be generated by institutions such as the World Bank, which, in the course of debt swapping, might dedicate new funds to the purchase of environmentally sounder technologies.

Finally, the United States, other developers of new technology, and international lending institutions should establish centers of training at locations around the world to create a core of environmentally educated planners and technicians, in order to "make the ground fertile" for sowing environmentally attractive technologies and practices--an effort not unlike that which produced agricultural research centers throughout the world during the Green Revolution.

With this SEI, we must transform science and technology to make it more efficient, consume less of the earth's natural resources, and emphasize waste minimization, recycling, and the use of renewable resources in harmony with the natural world. We must start by quickly obtaining massive quantities of information about the global processes now under way--through, for example, the Mission to Planet Earth program of the National Aeronautics and Space Administration.

And we must target first the most readily identifiable and correctable sources of environmental damage. For example, I have introduced a new comprehensive legislative package to effectively halt CFC, carbon tetrachloride, methyl chloroform, and halon emissions, and to promote development of technologies to replace those that now rely on CFCs.

Earlier this year, I introduced the World Environment Policy Act of 1989, a far-ranging bill to address virtually every aspect of the global environmental crisis. In order to accomplish our goal, we also must transform global politics, shifting from short-term concerns to long-term concerns, from conflict to cooperation.

Recent evidence leads me to believe that we have the capacity for this change. Just as the equilibrium of an environmental system can suddenly change from one state to another, so the equilibrium of one political system can suddenly change from one state to another. We politicians are frequently adept at symbolic action, a pretense of change without the substance of change. And for that reason my optimism is tempered by awareness of the power in the forces of greed and fear. But I do believe we have the capacity for what is needed--because the challenge can now be accurately described in terms of national security.

Some may believe that the idea of the environment as a national security issue is just rhetoric. Many of us, however, accept it as a statement of fact. But we also know that just as the world has been living with the possibility of man-made disaster in the form of nuclear war, so it now lives with the growing threat of man-made disaster in the form of catastrophic environmental failure.

In many ways, it is the same basic dilemma. In each case, our survival was threatened at a basic, primal level--the fear of death from attack by an enemy, the fear of death from running out of food. To each threat, we responded with more and more efficiency. The increasing sophistication of our technology has enabled us to confront each threat to our survival with a more powerful response. And in each case, the effort to secure our survival has instead threatened our survival.

Moreover, even if we are successful this time in meeting the needs of our survival and preserving the world's environment, it probably will not be the last time we will face this basic problem. Genetic engineering may pose the same dilemma all over again. In the effort to protect ourselves against disease, we are creating a new and more powerful technology that may ultimately confront us with the same historic challenge to human nature and the same hubristic relationship of our species to the limits nature has designed for us as part of the world ecological system.

As a result, it is hard to escape the conclusion that we must also transform ourselves--or at least the way we think about ourselves, our children, and our future. This last transformation is the most essential and yet the most difficult. If there is one cause for the prevailing pessimism about our ability to meet this unprecedented challenge, it is the belief by many that we are incapable of the change in thinking required.

And yet there are precedents that give cause for realistic hope. Human sacrifice and slavery were both once commonplace in human societies, yet both are now obsolete. Our thinking was transformed. These changes, like most changes in global climate patterns, took place over a long period of time. But now just as climate changes are telescoped into short periods of time, we must create in a single generation changes in human thinking of a magnitude comparable to the change that brought about the abolition of slavery. Yet once again, we must remind ourselves that the pattern of change required is visible only from a distance.

In this case, we cannot rely on science to give us a new point of view, for it is partly responsible for the problem. In ways not yet fully understood, the scientific revolution itself changed the way we saw ourselves in relation to the world. We detached ourselves from nature to examine the physical world. In a kind of Heisenberg principle writ large, we altered--without realizing it--the nature of what we began to examine. The new pattern of thinking we must now create is one in which we once again see ourselves as a part of the ecological system in which we live. What we now lack is a sense of the proper location of our species in the ecosystem. We have lost our "eco-librium."

How then can we gain sufficient distance from ourselves to see a pattern that contains ourselves in a larger context? My own religious faith teaches me that we are given dominion over the earth but that we also are required to be good stewards of the earth. If we witness the destruction of half the living species God put on this earth during our lifetime as a result of our actions, we will have failed in the responsibility of stewardship. Are those actions, because of their result, "evil"? The answer depends not on the everyday nature of the actions, but on our knowledge of their consequences. In an examination of Hitler's lieutenants, Hannah Arendt coined the memorable phrase "the banality of evil." The individual actions that collectively produce the world's environmental crisis are indeed banal when they are looked at one by one: the cutting of a tree, the air conditioning of a car.

"Evil" and "good" are terms not used frequently by politicians. But in my own view, this problem cannot be solved without reference to spiritual values found in every faith. For many scientists on the edge of new discoveries in cosmology and quantum physics, the reconciliation of science and religion sometimes seems near at hand. It is a reconciliation not unlike the one we seek between man and nature.

But even without defining the problem in religious terms, it is possible to conclude that the solutions we seek will be found in a new faith in the future of life on earth after our own, a faith in the future that justifies sacrifices in the present, a new moral courage to choose higher values in the conduct of human affairs, and a new reverence for absolute principles that can serve as guiding stars by which to map the future course of our species and our place within creation.

18

IMPLICATIONS FOR PUBLIC POLICY: OPTIONS FOR ACTION

Martin W. Holdgate

INTRODUCTION: A GLOBAL DILEMMA

The title of this forum begs a question: <u>Does</u> humanity <u>have</u> a common future in the face of global change?

We do not have a common present. We live in a world of depressing gulfs between rich and poor, educated and illiterate, cared for and neglected, and consumer and producer societies. We live in nations with dramatic differences in population growth and pressure on the natural resource base.

Lessons of history suggest that stresses are as likely to produce competition or isolationism as cooperation. Are not the stresses of higher global temperatures, shifting precipitation patterns, rising sea levels, and increased UV-B penetration more likely to exacerbate current problems, especially in regions whose population is likely to double within 25 years, and plunge nations into misery and strife? We already have environmental refugees moving from areas of depleted resources and causing frontier conflicts. What must be done if we are to prevent turmoil as a consequence of global change?

The answer, surely, is to try to get the most practical and effective policies followed where we can, as soon as we can, even if we cannot solve all the world's problems at a stroke. For being daunted by the challenge will solve nothing. Partial success is better than total failure.

THE CRITERIA FOR A SUCCESSFUL RESPONSE

Any public policy addressing this situation must meet five criteria:

1. It must start from the social and political realities of today's world, in all its diversity and complexity.

2. It must accept that governments and people have more immediate preoccupations than what the world will be like 50 years hence.

3. It must accept that it is the activities of the developed world that have largely caused the problems now confronting us and that the North is looked to by the South to solve those problems in a way that will not inhibit the South's essential development.

4. It must accept that policies and actions have to be developed nationally in ways that fit national traditions, politics, and aspirations and that are capable of effective implementation on the ground.

5. It must accept that national efforts must be linked internationally, but in an equitable way that does not impose the judgments of the North on the South.

I suggest that a successful global approach, a mosaic though it must be of international, national, and local actions, will have five ingredients:

1. Awareness building.
2. Avoidance strategies.
3. Adaptive strategies.
4. Abatement measures.
5. Assistance mechanisms.

These will be discussed in turn.

Awareness needs to be cultivated at international, national, and personal levels. There is evidence that governments are aware of the potential immensity of the problems: For example, the U.N. General Assembly responded positively to the debate initiated by the president of the Maldives, who pointed out that virtually none of his country's 1191 islands would remain after a 2-m rise in sea level. The U.N. Environment Programme (UNEP) has held informal discussions with ministers who placed global change high on their list of concerns. One hundred twenty-four delegations, 80 of them led at the ministerial level, attended the recent London Conference on the Ozone Layer. An Intergovernmental Panel on Climate Change has been established to consider the changes, their impacts, and responses. The Commonwealth has established an expert group to advise its heads of government on these same issues.

There is less evidence of effective national analysis and planning, and what there is lies mostly in the developed world. In the United States, Congress has requested that the Environmental Protection Agency prepare a report on policy options for stabilizing global climate. On April 26, 1989, the British prime minister, five members of the cabinet, and five other ministers discussed the questions with 50 leading scientists and industrialists. But these discussions are only slowly cascading down to catalyze strategies in individual firms and local communities. At a personal level, awareness depends immensely on the mass media. Such personal awareness is essential because global impact is the integral of innumerable human actions that appear trivial when considered alone. Twenty years ago, how could anyone who puffed an aerosol can have suspected that he or she was depleting stratospheric ozone over Antarctica in springtime? Even a woman in the Sahel who cuts the last bush in a desperate quest for firewood is likely to be aware of the local rather than the wider implications.

So building awareness is one key to the evolution of policy. Awareness-building campaigns should emphasize the need for sound science, including the gathering of more data for input to better models. They need to explain in nonhysterical terms what the scientific evaluations imply and what can be done. They need to build commitment at the citizen level to the expenditure, and to the changes in products on the market, that will be needed. In developing countries they will need to explain how local administrations can act to avoid problems and at the same time maintain the impetus of environmentally sound development.

Avoidance strategies and adaptive strategies are closely linked. The first means evaluating how the best available scenarios of global change could affect a region, a country, or an industrial corporation, and desisting from actions that could aggravate risk and loss. The second means changing plans and actions in a more definitive way to shift the process of planned investment and development onto a durable course.

For example, we have good reason to expect that sea levels will rise by some 10 to 20 cm by 2030 and will go on rising thereafter. It makes sense to avoid development low down on unprotected coastal plains and deltas that are at special risk. The U.S. Army Corps of Engineers calculated that natural salt marsh systems in Boston Harbor provided sea defenses that averted some U.S.$17 billion worth of expenditures on construction. Salt marshes, like mangroves in the tropics, have a capacity for upward growth if sediment is supplied, pollutant discharges from the land are controlled, and damaging overuse, like excessive cutting of mangroves, is avoided. Coral reefs, whose shelter is essential to the survival of many atoll nations, can grow at up to 10 mm per year if likewise protected from damage and pollution. While the capacity of all these systems may be outstripped if sea level rises follow some projections, it makes sense to survey such natural defenses and avoid action that reduces their effectiveness.

Forests, similarly, play a crucial part in controlling water runoff, preventing erosion, and in the Amazon, maintaining rainfall. The current destruction promises to establish a vicious cycle of climatic, hydrological, and production decline. Yet it is pursued for reasons that appear economically compelling in the short term. What is needed is the demonstration that sustainable use is more compelling. I have seen estimates that such sustainable extraction of timber, fruits, medicinal plants, and wildlife, correctly valued, can yield over U.S.$200 per hectare per annum as against a one-time U.S.$150 for irreversible destructive clearance. Avoidance strategies mean substituting more durable and resilient land use. Adaptive strategies mean preparing now for the changes likely to occur under the best available scenarios: reviewing cropping systems, forestry, river management, irrigation, and the distribution of urban and industrial development so that they will be in the right place at the right time. For those concerned with conserving biological diversity, adaptive strategies require looking at the pattern of protected areas, which in 50 years may no longer provide suitable habitats for the organisms they are supposed to safeguard.

We have to plan now, without waiting for models that give us a clear regional breakdown of likely change, because every decade we delay may aggravate losses. But because of the scientific uncertainties, plans themselves must be flexible and adaptive. One pattern we could follow is that of the National Conservation Strategy, adopted by many governments with help from the International Union for the Conservation of Nature and Natural Resources (IUCN).

A successful national strategy has three important features. First, it has to draw together the "horizontal" dimension of government, making all departments that depend on and influence environmental resources plan for their sustainable use, and this means involving finance, industry, energy, agriculture, forestry, and commerce as well as the obvious "green centers" of national parks and wildlife. Second, the strategy has to extend "vertically" from central through local government and has to bring in industrial corporations, village communities--especially women-- and citizen groups. It has to create genuine dialog about what to do for the common benefit. It has to promote a genuine attachment of the true cost of environmental change to development projects. And third, it has to create a continuing process, not just another document on the shelf. This kind of process, moreover, addresses the whole question of wise use of national resources; it therefore allows the impact of change to be put in a social context.

But while avoidance and adaptation can save lives, costs, and much pointless destruction, the lesson of this forum is that we face changes far too grave to allow us merely to adjust to them while their causes are unchecked. We face a bigger alteration of climate in a shorter time than the world's ecosystems have experienced even during the dramatic oscillations of the past 200,000 years. This will shift the zones of ecological tolerance of species hundreds of kilometers horizontally, and hundreds of meters vertically, with unpredictable consequences for natural systems and crops. Low-lying nations like Kiribati, Tuvalu, or the Maldives may face submergence in 100 years. We have to attack the causes and this means <u>abatement</u>.

Priorities for abatement need to be based on (1) the relative significance of a damaging substance and (2) the practicability of control.

Best estimates suggest that carbon dioxide (CO_2) contributes 50 percent of the greenhouse effect; methane, 18 percent; chlorofluorocarbons (CFCs), 14 percent; nitrous oxide, 6 percent; and others, notably tropospheric ozone, 12 percent.

There are special reasons to start with CFCs, also clearly incriminated as depleters of stratospheric ozone. The recent London conference concluded that we had to adopt as our objective the total elimination of production and consumption of CFCs and halons. It heard from industry that substitutes are available or are under development. It urged the universal acceptance, strengthening, and more rapid implementation of the Montreal Protocol. There is no reason why we cannot eliminate CFCs by the end of the century.

We can also press ahead with the application of existing technology to scrub sulfur oxides from flue gases and to cut emissions of nitrogen oxides from power stations and the precursors of tropospheric ozone from automobile emissions. All these things are relatively simple technically. If we cannot deal with them, there is not much hope for our action against more intractable greenhouse gases.

It may be more difficult to act against methane and nitrous oxide, since a high proportion of the former appears to originate in wet rice cultivation and livestock. But by common consent, abating CO_2 is the great challenge.

About 80 percent of anthropogenic CO_2 emissions come from fossil fuel combustion. Coal, oil, and natural gas all yield it, but coal produces the most per unit of energy produced. North America contributes 27.5 percent; Eastern Europe, 25 percent; Western Europe, 15 percent; Asia, 8 percent; the Pacific region, 6 percent; and the other developing countries, 5 percent; so it is very much a developed world responsibility even though the developing countries bid to overtake us in 30 years. As a matter of equity, it is the developed countries that are looked to for abatements that will make room for Third World growth without environmental disaster.

One target might be to halve CO_2 emissions from the developed countries by 2020. This is ambitious, for current energy-use projections imply an increase of over 25 percent on that time scale. I have seen reviews suggesting eight possible approaches:

1. Reforestation.
2. Generating energy from organic waste, which otherwise goes to landfills and releases methane.
3. Improved efficiency in the generation and use of energy from fossil fuels.
4. Substitution of fuels like natural gas that produce less CO_2 per unit of energy generated.
5. Removal of CO_2 from power station flue gases.
6. More efficient use of fuel for transport.
7. Development of renewable sources (hydro, wind, wave, geothermal, solar, and fuels from biomass).
8. Increased use of nuclear energy.

The optimal mix and scale of benefit from each must vary from country to country. Some, like reforestation and generation of energy from waste, are worth doing and can have great public appeal, even if their contribution is small. Others, like substituting natural gas for coal, presuppose the availability of supplies. In most developed countries, however, the most promising approaches are likely to be energy conservation (which in Europe could contribute 40 percent of the target), development of fuel-efficient transport (12 percent), and if the public can be satisfied on grounds of safety, increased use of nuclear energy (23 percent).

The target is attainable--at a cost. But only if the citizen is prepared to act. For energy conservation in the home, factory, and office depends on consumer commitment, and economies in transport may mean some reduction in vehicle performance. Market forces can help both, and this points to an expensive energy policy with, for example, the threatened tax on gasoline of U.S.$1 to U.S.$2 per gallon. Tax incentives can help pull through investment in home insulation. Public statements of technical targets may be needed to bring forward more efficient products like the 80-miles-per-gallon automobile (recall that technical targets for emission abatement pulled through the 3-way catalyst). There is, self-evidently, a feedback loop here to awareness, which can pull products through by creating environmentally sound markets.

Internationally, if not nationally, <u>assistance</u> will undoubtedly be needed to secure the changes required. Many of the poorest countries are also among the most vulnerable to climatic change, if deserts become more arid and the sea encroaches on densely populated coasts (and half the population of the world lives in coastal zones). Such assistance will be needed for three main purposes:

    1. To finance national surveys and development of avoidance and adaptive strategies.
    2. To promote development of forestry and agroforestry systems and other land-use practices most likely to be resilient in the face of change.
    3. To transfer technology that will incorporate substitutes for CFCs and greenhouse gas abatement methods.

Cooperation will also be needed in ways that go beyond traditional development aid. World markets and trading patterns must be adjusted to favor products from developing countries suited to a changing world. Real resources will need to be transferred from the North to the South to promote the conservation of biological diversity that the North uses but the South shields. We need also cooperation in science to gather good worldwide data and develop better models. A good start has been made through the World Climate Program, the Intergovernmental Panel on Climate Change, and other groups.

    Beyond this, there is a case for expressing our common commitment in a framework convention on the limitation of climatic change. This might bind its contracting parties as follows:

    1. To observe a code of conduct in respect of measures to protect the atmosphere.
    2. To cooperate in research and assessment.
    3. To provide assistance for technology transfer.
    4. To negotiate specific protocols for the limitation and reduction of greenhouse gases.

Such a convention will inevitably take thorough debate, but it is my impression that there is international recognition of the need to do something to give voice to the collective commitment of the world community.

    Finally, we need to review the world's institutional machinery. Primarily, this means the United Nations. There have been many proposals, but I suggest it will be easiest to work with the structure we have.

    1. The U.N. Security Council, whose mandate is wide enough, might periodically review major environmental issues of global concern that could threaten peace and security.
    2. A commission on environmental change might provide technical backup to the Security Council, perhaps continuing the work of the Intergovernmental Panel on Climate Change.

3. The Agencies Coordinating Council could make a real effort to harness the U.N. agencies together.

4. The U.N. Environment Programme, perhaps given a new mandate at the second U.N. conference on the environment due in 1992, could continue to provide a wider forum for environmental discussion.

But the United Nations alone is not enough. The action pattern we need is much broader. As chief executive of the world's largest professional organization concerned with conservation, may I make a plea for the nongovernmental organizations (NGOs)? With a membership of 62 governments, 130 state agencies, and over 300 NGOs active in 120 countries, we can operate with a flexibility denied to the intergovernmental machinery and build institutions especially in the developing world. Such machinery is needed, for the developing countries need a vehicle for stating their needs and reversing the traditional dominance of "northern" concepts and priorities.

CONCLUSIONS

My conclusions are simple. We must stimulate public awareness of the problems that confront us, but on the basis of good science rather than exaggeration and half-truth. We must promote individual commitment to actions that may cost more now, to save immense costs later. We need to promote national, local, and corporate measures of avoidance and adaptation, and we should do this now in parallel with the increased perfection of science, which we also need. We must accept the need for assistance, especially to the developing world, if they are to pursue energy-efficient pathways. And we must redesign our international machinery for coordination and cooperation, including the development of a new international instrument or convention and stronger organizational mechanisms. We may not succeed everywhere--I admit to being pessimistic about the prospects in some areas--but we can ensure more success by acting now than by doing nothing in the misplaced hope that the problems will solve themselves.

19

VIEW FROM THE NORTH

Digby J. McLaren

This discussion does not represent the view from the North but merely a view from the North. ("North" in this title means Canada.) Furthermore, although I am interested in and concerned about the questions of global change and Canada's part in the world program, I speak officially for no government or organization within my country. I did not have time to assess the official view, and indeed when I do know what it is, I am not necessarily always in agreement with it.

I can, however, assure you that there is strong support from working scientists for the research program on global change. Already a great deal of good science is being carried out in a very large spectrum of disciplines that might be considered grist to the program's mill. The scientists that are already involved in such activities might not necessarily be aware of their contribution to the program, and there is a huge job to be done in overcoming generations of traditional specialization in science disciplines. We are now required to learn how to communicate broadly with people in other disciplines and indeed with those in the social sciences and humanities, the decision makers, and, more important than any other, the public. The Royal Society of Canada was involved in the planning of a global change program from the inception of the idea, and it set up a research committee in 1985, which subsequently organized working groups geographically and by discipline. Canada early recognized the importance of collaboration with the "other cultures" and unified all in one program, subdivided into two overlapping parts--the human dimension and the scientific dimension.

In the world program, Canada has a double obligation. First, it is the second largest country with the longest coastline and must therefore play a major part in contributing to the whole. Plainly there must be a major effort in the Arctic, a region particularly sensitive to change; in agriculture, which is found only in a small southern strip currently restricted by climate and whose future expansion is limited by geology; in water and wetlands and their huge importance in moderating atmospheric chemistry and the effects of northern climate as well as their sensitivity to climate change; and many other ecological concerns. In the oceans, Canada plays its part in the major international programs, e.g., the Ocean Drilling Program, the World Ocean Circulation Experiment, and the Joint Global Ocean Flux Study. Second, as a "have" nation, Canada has shown a strong interest in the Third World through existing

195

organizations. Attention has been given to tropical rain forests and a number of geographically important areas: Southeast Asia (Borneo, Thailand, Bangladesh, Nepal), Africa (southern Africa and the Sahel), and Latin America (the Pacific coast of South America).

The most important aspect of the research program on global change is that it is truly global and must involve all people on earth. To point this up, Bill Fyfe has recently suggested that the developed countries, because they are outnumbered four to one by the lesser-developed nations, should pitch their level of effort to populated areas containing four times as many people as their own population. Canada should, therefore, accept responsibility for collaborating with about 100 million people, at a minimum.

The Brundtland Commission has provided us with a baseline on which to build. This hugely important work was the subject of two major meetings in Canada, and the results were published as the <u>Brundtland Challenge and the Cost of Inaction</u>, involving scientists, economists, moralists, and politicians. Canadian granting councils and others are considering proposals for major projects on economic, urban, agricultural, and other dimensions of the social challenge of global change. The Brundtland publication <u>Our Common Future</u> (World Commission on Environment and Development, Oxford University Press, London, 1987) coined the term "sustainable development." The commission was content to link sustainability with discipline and restraint but suggested that growth might continue. While in no way critical of this classic work, I should like to examine some aspects of sustainability.

o <u>Sustainable for whom</u>? It appears that our economic system emphasizes short-term profit as a benefit and does not put a real cost on the resources we consume. There must be a price put on such commodities as soils as well as ground water, surface waters, atmosphere, and the biosphere--or the sum total of all kinds of life on earth. Current economic thinking appears to be caught up in a system that assumes limitless resources and ignores the production of waste products. This system worked when resources did appear to be limitless and when waste was easily disposed of and self-cleansing. Neither of these qualities exists any longer. The economic subsystem takes in resources and excretes waste and is thus irrevocably and closely linked to the ecosystem. Input and output are finite, and the main variable is the one-way flow of matter-energy. Such a way of looking at things raises the question of how big the economic system should be in relation to the physical dimensions of the global system. This also necessarily questions the concept of growth economics and the impossibility of generalizing western standards to the world as a whole. Since one-quarter of the world population uses most of the resources and produces most of the waste, can we increase both in the other three-quarters? What are the limits that must exist in every finite system?

o <u>Sustainable until when</u>? The population of the world has fluctuated widely in the past, controlled by the harsh laws of the natural environment--flood, famine, plague, and conquest--and it has remained well below 1 billion until the Industrial Revolution. With the present population at 5 billion, therefore, we may claim that this runaway growth is the direct result of the application of technology, primarily improved food production and hygiene, and that it is not a product of the natural environment. The increase in the present population is grossly unequal geographically. The largest birth rates are still in underdeveloped countries, but in regard to resource use and production of waste the inequality is reversed and taken over by the developed world. Currently we are adding 1 billion human beings every 11 years. There just is not enough time for demographic theories of population stability, arising from increased standards of living and education, to work. Recent global projections suggest stabilization at 11 billion by the end of the next century, which approximates the date when, at current rates of destruction, all forests will have been felled. How will all these people be fed? Any global attack on the disequilibrium of earth's systems must take these facts into account.

    o Sustainable energy. The present system of energy use faces an inevitable change. The fossil fuels currently account for about 79 percent of world usage of energy, and 72 percent of this is oil and gas. Availability will dictate a reduction in the use of oil and gas, which do not have an unlimited future. Coal represents at least 10 times the stored energy of petroleum products. Burning coal also produces nearly twice as much carbon dioxide per unit of energy produced as does natural gas, along with a large number of highly undesirable other wastes. It has been suggested that, if one-third of the total global coal resource is used, the atmosphere may pass the point of no return and become subject to a runaway greenhouse effect. The warming effect appears to have started: How much shall we accept, bearing in mind the inevitable change in climate and rise in sea level? When shall we make the change-over? How shall we pay for it? Do we have the technology ready?

    o Sustainable ecology. Humankind lives inside the environment and, in spite of biblical license, cannot dominate the whole of life on earth. We have heard how the biodiversity of the planet is being reduced by many of our current agricultural and other exploitative practices, and we appear to be entering a major extinction event. There have been many major events in the past during which the planet lost 80 percent or even up to 90 percent of the extant biomass at the time of the catastrophe. Without going into the question of what causes such mass extinction events, one can make certain observations from the past, which should serve us today. Mass killings have been sudden, but this just means that they cannot be resolved accurately at this distance in time. The current extinction event may be as sudden as any of them. When the killing is over, it takes a very long time to reestablish ecological equilibrium. Certain major features of the ecology of the earth are reestablished with difficulty, and new ecologies must be worked out by a long, slow process of selection. Thus coral reefs have become extinct at many times in the past, but their recovery has commonly taken as long as 10 million years. Other animal groups have shown a similar pattern in the past.

    This talk is not meant to be preaching doom, but I do not believe that describing reality, or the best interpretation of what reality is from the facts given to us, is undesirable. I am not a pessimist; I believe that humankind will survive the present crisis, but it will not do so unless it faces the reality of what is going on in this small planet. We cannot dominate life. We do not understand the intricacies of earth ecology; to pretend that we do and to act accordingly are to court disaster.

Perhaps a way may be found to judge our actions by a new principle: the health of the planet. The economy could be seen as being within the environment and not the environment as being within the economy. Perhaps we should ask the question of any action that we take, Does it increase or decrease survivability? The term "quality of life" might take on a new meaning. We could try to limit our own needs with a consciousness of the global resource and global needs in the broadest terms. We could live in balance with all life but could crop and harvest according to preference and needs, providing that these do not reduce the capacity of future generations to do so also.

In a similar way, we may look on sovereignty in a different light. We may assume a common cause with all people on earth against a common enemy--action that threatens balance within our environment or reduces our legacy for future generations. Somehow a way must be found to permit us to look for one brief moment at the world without the filters of be- lief, axiom, or political theory. In this moment we could observe the planet and draw conclusions from our observations as to the health of our habitat and assess the probability that life may be self-sustaining indefinitely into the future. Sustainability is the ultimate criterion by which we must measure our behavior and influence the presuppositions that lie behind all our beliefs.

20

VIEW FROM THE SOUTH

Marc J. Dourojeanni

I am here at this forum as a Latin American, and most of my comments and examples refer to that part of the world. Of course, the concepts I develop on this occasion are personal and do not represent, in any form, the views of the World Bank. Also note, please, that when I refer to "the North," it indicates all developed countries, including socialist developed countries, not just the United States of America. When speaking of "the South," I mean all Third World countries, less-developed countries, or developing countries.

My presentation is organized in two parts. In the first I explain how policymakers in the South perceive the North's concern for environmental affairs as related to southern countries, and I outline the elements the South considers necessary to establish a productive dialog. The second part deals with what are mainly southern responsibilities, which most policymakers recognize as needed. Of course, the views on this matter are no more uniform in the South than they are in the North. Nevertheless, there are trends that I will try to describe.

REQUIRED CHANGES IN NORTHERN POLICIES TO CHANGE
SOUTHERN PUBLIC POLICY TO MEET THE GLOBAL CHALLENGE

1. <u>The first question is, Are we seeking planetary security? If so, the North should eliminate the risk of nuclear war</u>.
The three most serious global threats for mankind are, first, nuclear war; second, regional and local conflicts; and third, the environment at large, beginning with the population issue. The risks of a nuclear war, today relegated to the back of people's minds, are still omnipresent. In the event of a nuclear war, although the South will suffer longer, it too will eventually be annihilated. The South expects the North to do much more to eliminate the risks of a nuclear war before blaming others for comparatively lesser risks to mankind and civilization.
2. <u>The North should practice what it preaches and should be more serious about its own contradictions with regard to environment, before attempting to rule the international environment</u>.
Politicians as well as common people in the South perceive the North as lecturing others but not following its own recommendations: telling

the South to save energy instead of developing new sources while wasting most of the planet's energy; opposing the construction of nuclear plants while building more plants and peddling nuclear technology. Blaming the South for what is mainly a northern responsibility, such as ozone layer depletion and global warming, is also perceived as an irony.

Of course, it is not possible to solve every northern environmental inconsistency before addressing the South's problems. If so, it will be too late for mankind. But more humility is required for a constructive dialog. Earlier, this year, two South American ships had minor accidents in the Antarctic. The North bitterly blamed the South for "ecological disasters." When, a few weeks later, the Exxon Valdez spilled oil in Alaska, causing the greatest ecological disaster since the Torrey Canyon oil spill in 1967, the people in the South were relieved to see that the North, as usual, had done worse than the South. This is a very bad sentiment, I agree.

As a matter of fact, the North cannot escape from its direct and indirect responsibilities in the environmental situation of the South. The current situation of developing countries is, in general, a consequence of past colonialism, neocolonialism, and imperialism. However, because we realize that we share a common future, developing countries are urged to guarantee also the safety of the developed world. The press in the South is flooded with reports and comments reflecting these views. They also underline that much environmental mismanagement in the South is a consequence of inventions, technologies, and "panacea" solutions of the North. Who invented and promoted massive utilization of agrochemicals? Who invented and promoted all the paraphernalia required to waste fossil fuels? Who consumes drugs and makes the lion's share of profits with their trafficking? Where is the market for tropical hardwoods, endangered wildlife, or meat produced in cattle ranches in the tropics? Who in the South promotes the wasteful consumption habits from the North's lifestyle? All these questions, and many others so often asked, are constantly present in the minds of southern politicians.

In addition, southern politicians witness another kind of contradiction. They receive almost simultaneously groups of foreign visitors offering lucrative business in tropical hardwood, and others from the same country demanding urgent action to stop this trade. Contradictions are normal and part of the democratic game and associated freedom. Southern politicians recognize that, but they want equal tolerance with regard to their own national contradictions. The North should give more consideration, not necessarily final solutions, to its own contradictions for the development of a sincere and productive dialog about global environmental issues.

3. <u>The North should abolish the concept of "donor" with regard to environment and to everything else in its relations with the South</u>.

Among the terms and related concepts that poison the international atmosphere, "donor" is the worst. Even the international finance agencies consider themselves as "donors" despite the fact that their loans are a part of the South's external debt. A recent document about U.S. cooperation for international growth and development rightly said that it is necessary to shift from the old concept of "aid" to the new idea of mutual gain through cooperation, but the document used the word "donor" everywhere else in the text. The future of North-South relations must be based on equity, mutual collaboration, and cooperation--on partnership in a common task.

4. <u>Foreign debt: Many Southerners claim that the North is as responsible as the South and that global security will be increased by an equitable solution</u>.

The South recognizes its own responsibility in the growth of its debt, but current generations simply do not have the ability to pay it. The accumulation of mistakes of older generations or, more precisely, from traditional oligarchies and dictators, is too heavy to allow any environmental solution.

5. <u>Both controlling the demand of "dangerous" natural products in the North and establishing fair prices for them are key elements for arriving at a global solution</u>.

The South will hardly be able to practice conservation as related to goods for which the northern demand is practically unrestrained, as in the case of drugs. This is the case also for other goods directly associated with environmental degradation, such as tropical hardwoods, meat (the "hamburger connection"), fish meal for livestock, pets, and the tragic trade in laboratory animals. Simultaneously, fair prices are necessary in order to pay the cost of management for sustainable production, but fair prices rarely exist. Tropical hardwoods are a good example. Concerted action between southern producers and northern consumers is necessary.

6. <u>A code of ethics for multinational business is urgently required</u>.

Who sells perverse development options and dirty technologies to the South? The criterion that any business is good if it is economically profitable should be reviewed. Good business should be much more than money. The "Bhopals" of history should disappear forever.

7. <u>Some respect and recognition for the efforts of the South to improve the environment are necessary elements of the future dialog</u>.

Southern politicians note that the South is doing proportionally more than the North to conserve nature and natural resources: Note the thousands of new protected areas and new nongovernmental organizations, innovative new legislation, public administration reorganizations, and so forth. The money invested by the South in environmental matters is several times more than any amount invested there by northern "donors," but there is little or no recognition of these efforts. On the contrary, as in the case of Brazil--a country investing more than a billion dollars in the next few years to conserve nature and improve the environment-- there is permanent negation by the North of such great efforts.

8. <u>Take it easy, do not rush. Local policymakers need time to digest the new environmental concepts northern experts produce so quickly, often giving new names to old concepts</u>.

During the 1960s the North promoted the establishment of "strictly protected" areas. During the 1970s the relaxed concept of "biosphere reserve" was launched. Now in the 1980s everyone has been proposing "sustainable development," even inside national parks, and so on in any field of the environment. Not even southern scientists have the capacity to follow all the new terms, concepts, and "scientific fashions." Also, northern experts could be perceived as arrogant when, after a couple of short visits, they know better than any national citizen what is needed. In this regard, remember that the task to assist in the development of a country requires a deep knowledge of the local reality and much humility.

9. <u>Finally, the United Nations is the best forum to deal with our common future</u>.

The United Nations is the best guarantee for developing countries that their views will be listened to and discussed. All global planning about our common future that is done in other forums, avoiding direct confrontation with southern politicians, might be successful for a while but in the long term is bound to face irreversible problems. But, of course, the United Nations system should be reorganized, strengthened, and better utilized.

SOUTHERN NATIONAL POLICIES REQUIRED TO MAKE THE GLOBAL ENVIRONMENT SAFER

1. <u>The first task of southern governments is to give higher consideration to the population growth factor</u>.

Any intention to ameliorate the environmental quality of life has been and will continue to be overwhelmed by rapid population growth. Most developing countries have sound family planning. Nevertheless, what they achieve is far from what is needed. Meanwhile, the Catholic Church and some powerful groups fight almost every realistic form of birth control. If the antiabortionists win the ongoing battle in the North, what is going to happen in the South? The most probable scenario is that, thanks to the influence of local versions of these groups funded by the North, the situation will worsen. It is imperative to deal openly and seriously with this problem, which is at the root of poverty and inequity and of every major environmental problem.

2. <u>Education, education, and more education is the oldest answer and still the key for most southern environmental problems, including family planning</u>.

Education at primary, secondary, and university levels, as well as public education and information, despite the immensity of the task, is the principal means to change human behavior. While education in the South should be renewed, improved, and expanded, almost every indicator of educational quality in the South shows that it is deteriorating.

3. <u>The South should dramatically reduce its military expenses to allocate more funds to internal security, including education, the environment, and food security</u>.

Politicians as well as many military establishments in the South recognize this necessity. But international conditions to allow such a change are not in place. The main effort should be undertaken at the regional level. But, in fact, very little is being done. Worldwide consideration should be given to reducing military expenditures and putting more emphasis on all forms of internal security. Anyhow, terrorism and guerrillas cannot be controlled with submarines or aircraft carriers. And environmental abuse exacerbates insecurity, through refugees, hunger, and poverty. Security can be achieved more economically by environmental management than with armaments.

4. <u>Debt swaps as well as other imaginative solutions to the debt crisis are becoming increasingly more attractive for the South</u>.

The few rejections of debt-for-nature swaps were, in substance, a consequence of the way they were submitted, particularly with regard to the sovereignty issue. Again, some time is necessary to make politicians aware of the new option. I believe that one of the main difficulties with the concept of debt swaps is currently its small scale, which makes it unattractive to countries with large foreign debts. The concept of a debt swap should be more aggressively expanded to include education, population, health, terrorism control, drug control, and many other items of international concern.

5. <u>The financing of forest protection, ecosystem reconstruction, forest plantations, and forest management in the South to compensate for northern pollution (e.g., "carbon sink" forests) is a good idea that should be supported</u>.

In many southern countries there is land available for the above mentioned objectives, which are fully justified in social, economic, and ecologic terms. The main limitation to carrying out these activities is the lack of long-term funding. This initiative will certainly be well received in most countries if the conditions are acceptable. The North should be ready, if this program is recognized as important, for more serious commercial competition from the South in forest products.

6. <u>The South is ready to delegate many governmental functions to the regional and local governments, ensuring a more careful treatment of environmental issues by citizens directly suffering the consequences of careless development</u>.

Decentralization and deconcentration are going on in almost every southern country. The first result of this trend is a much more serious treatment of environmental issues. The politicians of the South are in favor of a faster move to regional and local enforcement of environmental legislation. This is one of the most important positive changes of the last decade.

7. <u>The necessity of limiting the growth of the megalopolis is recognized in the South, although the concrete measures to achieve this goal are not yet clear</u>.

The urban environment is becoming one of the most challenging problems for the South. Urban planning is poor, especially in the long term, and it is a direct consequence of problems occurring elsewhere in the country. The promotion of the development of mid-size cities should be seriously considered as an alternative.

8. <u>All southern politicians also recognize that the lack of good long-term planning is one of the major causes of environmental degradation</u>.

Little is indeed being done about this problem. Plans exist, but they are rarely followed, even by those who submitted them. Development projects, because of this lack of overall planning, frequently are of little benefit.

9. <u>Social reforms, such as land tenure reform and other equivalent tools to establish the base of equity, are a requisite for a common future</u>.

The need for social reforms is accepted, but actions to achieve them are currently limited. Part of the problem is that right-wing politicians, who are often more interested in the environment than are left-wing politicians, seem unable to correlate problems such as deforestation caused by landless farmers with large estates producing export industrial crops.

10. <u>Finally, the South should revise its economic policies, especially those that offer economic incentives for environmentally destructive activities</u>.

Taxes, subsidies, and credits, as well as regulations, should be studied and changed if they cause negative impacts on natural resources and the environment. An important way to achieve this may be provided by the valuation of natural resources and its incorporation into national accounting.

FINAL REMARK

Equity is the basic requisite for global environmental security. Southern countries are still far from achieving justice and equal opportunities for all. Many deeply required social reforms, such as land redistribution, are still not always being considered essential steps for peace and for a safer environment. On the other hand, the North should remember that equity is as required in North-South relationships as it is inside southern countries.

21

POLITICAL LEADERSHIP AND THE BRUNDTLAND REPORT: WHAT ARE THE IMPLICATIONS FOR PUBLIC POLICY?

Charles Caccia

It came as a shock in March 1989 to learn from an Environmental Protection Agency (EPA) report that about 1 billion kg of toxic chemicals are released annually and 100 million Americans breathe pollutants that exceed federal standards. Representative Henry Waxman was quoted as saying that "EPA has broken commitment after commitment to deal with this problem" during the 19-year life of the Clean Air Act. In Canada, we are hardly innocent--the Canadian record on water pollution is not much better. We are better off on air pollution, mostly because we have more air! When one adds the mostly ignored principles of the Stockholm conference and the near-paralysis of the Law of the Sea, the inevitable question is, What is going on? What kind of game are we engaged in? Are the implications of such behavior relevant to the implementation of the Brundtland report (World Commission on Environment and Development, Our Common Future, Oxford University Press, London, 1987)?

We seem superbly gifted in articulating principles, as proven by the Stockholm Declaration. We also know how to define the problem and even how to prescribe remedial action. Some political leaders have even embraced sustainable development: Prime Minister Margaret Thatcher in July, President George Bush on August 31, and Prime Minister Brian Mulroney in September, 1988. But to what extent do they mean it? Do they know what they are endorsing? Will a public, often prone to forgetfulness, hold them accountable in the polling booth? Politicians have been known in the past for drafting and even passing legislation yet paralyzing its enforcement by denying adequate resources, or even failing to proclaim approved legislation into law, as in the case of the Vehicle Fuel Consumption Act, passed by Parliament in Canada in 1982 but never proclaimed into law.

Our judgment of governments should not be entirely determined by their performance with respect to the Stockholm Declaration, The Law of the Sea, and the U.S. Clean Air Act. But there are good reasons for being skeptical. Principle 21 of the Stockholm Declaration reads: "States have, in accordance with the charter of the United Nations and the principles of international law, the sovereign right to exploit their own resources pursuant to their own environmental policies, and the responsibility to ensure that activities within their jurisdiction or control do not cause damage to the environment of other states or of areas beyond the limits of national jurisdiction."

Yet 17 years later, signatories of the Stockholm Declaration continue to dump their acid gas pollution on each other. The United States, the United Kingdom, and Poland, to name three major industrial nations, have so far ignored the Helsinki Protocol on sulfur dioxide reductions. Take the Clean Air Act: Why must Representative Waxman scream 19 years later about lack of enforcement? In the case of the Law of the Sea, why have Canada, the United States, the United Kingdom, and the Soviet Union, to name a few major maritime nations, not signed or ratified it? In the search for explanations for this strange behavior, where should we look?

Science, a human enterprise based on skepticism and caution, seems to understand and agree on the urgency of global change on many fronts. The public--with the help of the media--accepts the need for action on global change. Politicians, at least individually, seem to understand global change to the point of saying the right things. Almost everybody seems to agree that the situation is both real and urgent. Why, then, are governments so reluctant to implement, to act, to enforce? Why not practice what they preach? Having endorsed the implementation of the Brundtland report with a mile-long resolution at the U.N. General Assembly on December 11, 1987, why are the major nations of the world not even beginning to restructure any part of their economies? We had best find out soon.

Part of the answer is to be found in attitudes slow to change at the deeper level necessary to alter both personal and institutional behavior. Where and what are the triggers required to bring about those deeper attitudinal changes? How do we move from worry to action?

How do we change attitudes that underlie the following?

o Rampant consumerism and manufacturing for planned obsolescence.

o The persistent shifting of costs of production on to water, air, and soil.

o The undue emphasis on the offensive approach to defense policies.

o The heavy drawing on the earth's capital, particularly in forests, fisheries, ground water, and top soil.

o The tendency to see the earth, as David Barash puts it, ". . . as something to be conquered rather than to be appreciated, as challenging and threatening rather than nourishing and protecting."

True, some new attitudes have emerged, here and there, and have even reached the level of institutional thinking in some political parties and governments. But just at the margin, the edge. The debate has just started in search, for instance, of ways of integrating the economy with the environment. But substantive decisions, the tough and necessary decisions, are hard to find. We are still toying with demonstration projects and "maybe someday" commitments. For example, take present-day political leaders in North America and the United Kingdom who have endorsed the Brundtland report: Are they just now waking up, or are their basic values and ideologies fundamentally incompatible with Brundtland's concept of environmentally sustainable development? Or both? We all have a good deal of rethinking to do.

Having embraced environmentally sustainable development, how do political leaders propose to proceed? Can contemporary conservatives become regulators and use government in a restructured form to implement Brundtland's blueprint? And if they cannot, since they believe in market forces, are conservative leaders prepared to really use the marketplace to achieve sustainable development, rather than using the marketplace as an excuse for inaction? Will they offer real incentives to nonpolluters? Will they impose heavy taxes on polluters?

If Mulroney, Bush, and Thatcher mean what they say, the marketplace is there for them to use, in keeping with their ideology. Yet their tax systems remain, for the most part, unchanged, blissfully ignoring incentives and obstacles to environmentally sustainable development. Their energy policies have drifted back to pre-oil-shock days. Their agricultural policies still rely on heavy use of fertilizers and pesticides, with consequences for ground water and health. Forests are being depleted. Fisheries are in danger because of overharvesting, not to mention oil and other toxic chemical spills. Development aid programs are mostly in a pre-Brundtland mold.

Mind you, it is not all discouraging news since the Brundtland report. The government of Norway is in the lead with the adoption of their white paper incorporating sustainable development in the policies, budgets, and programs of each and every department of government. In The Netherlands, an action program is scheduled to be presented to parliament any time. In Japan a white paper was produced in response to the Brundtland report, outlining Japan's contribution toward the conservation of the global environment. In Indonesia the government approaches sustainable development in its seventh national economic plan. In Canada the Task Force on Environment and Economy reported in October 1987. Roundtables on sustainable development have been announced and formed.

The fact is that in the major industrial countries, while the public seems ready to accept stiff medicine to ensure a future, governments' actions seem no more than sporadic. Governments are still lurching from catastrophe to catastrophe. They are seeking refuge in declarations, most recently in The Hague on March 12, 1989. Between Saint-Basile-le-Grand (site of a major PCB fire in Canada) and the Exxon Valdez spill in Alaska, we go back to sleep until rudely awakened by the next environmental disaster.

Missing is the leadership that can translate emerging values into tough but necessary decisions that would change trends. We remain reactive rather than proactive. Yet public opinion seems ready to accept leadership capable of giving the creative momentum required to change trends. But we still seem trapped. Why?

Why is it that we seem to be behaving like institutionally bound lemmings? We have a public well ahead of most decisionmakers in matters affecting global security, be it the environment or defense. All over the world, the public wants peace and security, and safe and healthy living conditions. People want environmental integrity for their own sake and the sake of their children and grandchildren. But governments, particularly governments of large industrial nations that could act on the publicly expressed will, are still making declarations. Why?

Is it because sectoral economic interests stand in the way? Is it because the common interest gets lost in a welter of special interests? Is it because calcified policies pose seemingly insurmountable obstacles to implementing a new agenda? Is it because public pressure is still smaller than the pressure exerted on decisionmakers by vested interests? Yes, all of those. It is these barriers that must be cleared. And quickly.

Specifically, what is it that needs to be done?

o Replace the present structure of government, in which the environment operates in isolation and in competition with other departments, with a structure in which the environment becomes the responsibility of all departments and agencies.

o Demonstrate that social and economic good can be achieved through environmental action, with resulting jobs in conservation, new industries in pollution-control technology, rehabilitation of water, air, and soil quality, improved public health, and better quality of urban life and planning.

o Make environmental impact assessments of major economic sectoral policies mandatory before mega-projects are approved and major decisions are made.

o Ensure that we live off the earth's interest without encroaching on its capital, investing to sustain and even enhance that capital so that future dividends can be ensured and enlarged.

o Make the promotion of energy efficiency a high priority and reduce our dependence on nonsustainable and environmentally risky energy sources.

o Reform tax systems with a focus on prevention of damage to air, water, and other resources.

o Provide incentives and support to the most energy-efficient modes of transportation.

o Include the scientific community in regular consultations with governments.

o Reduce agriculture's dependence on chemical fertilizers and pesticides so as to restore surface and ground water quality.

o Encourage and adopt production processes and technologies that do not damage the environment.

o Reduce international debt levels by using funds made available from a reduction of armament expenditures.

o Sign and ratify existing international laws and protocols affecting the environment.

o Give priority to the preparation and implementation of a global convention on the protection of the atmosphere.

o Establish international standards of behavior in the global commons.

o Urge the U.N. secretary general to constitute a special U.N. Board for Sustainable Development and the U.N. General Assembly to create a Security Council for the Environment, to which such a board would report.

To say that we need all this and more to happen is not enough. We must find ways to make it happen. The key to making it happen lies in the values, the vision, and the creative capacity of new political leadership.

The nature and speed of change, the complexity of issues, the inertia of institutions, and the power of vested interests are such that they require leadership qualities that usually emerge in troubled times, with the hallmark of rendering difficult decisions acceptable--leadership that can mobilize the public because the public senses integrity, commitment, and foresight; leadership that believes in the vital role of government in regulation, restructuring, advocacy, and education. Declarations alone do not change trends. Exhortations to proper behavior are just exhortations. They are not very effective. They offer good escape hatches, good public relations for the short term, but they are symptoms of impotence. Creative leadership is what is needed to implement Brundtland's prescription. In the past, Pierre Trudeau, John F. Kennedy, and Franklin D. Roosevelt were able to bring about widespread acceptance of difficult decisions. They possessed creative intellects. They overcame obstacles. They brought about change. They knew how to use government, democratic government, trusted government, and all its instruments for the collective good.

For global change is a matter of collective interest and survival. We must find again ways of generating the leadership required at all levels of government, in all sectors of society, to change present trends as Gro Harlem Brundtland has urged. And for a simple reason, given so poignantly by way of a graffito painted on a bridge in Rock Creek Park in Washington, D.C. It reads: "Good planets are hard to find!"

22

GLOBAL WARMING: IS IT REAL AND
SHOULD IT BE PART OF A GLOBAL CHANGE PROGRAM?*

Stephen H. Schneider

The summer of 1988 saw a combination of events that quite literally catapulted the "greenhouse effect" out of the halls of academe and government offices and into the public consciousness. Major drought, intense heat waves, forest fires, a super hurricane, and flooding in Bangladesh--all the kinds of events that had long been forecast as associated with global warming from increasing greenhouse gas build-up--occurred in that year. For months, magazine covers, news broad-casts, and newspapers were dominated by stories of the extreme weather. These were often juxtaposed with stories on the greenhouse effect. It would have been hard for a listener or reader who was not scientifically trained to conclude anything other than that there was an intimate con-nection between the bad weather of 1988 and the buildup of greenhouse gases that had been increasingly discussed by a number of scientists for more than 15 years.[1] Was this finally the proof that these academic warnings had indeed come true?

On June 23, 1988, James Hansen of the National Aeronautics and Space Administration's Goddard Institute for Space Studies (GISS) stated be-fore the Senate Committee on Energy and Natural Resources that he con-sidered there was a "99 percent" chance that the unusually warm globally averaged temperature records he and a colleague had constructed for the 1980s could not have occurred by chance but rather were the result of the buildup of greenhouse gases.[2] He went on to point out that increasing the global temperature would increase the likelihood of extreme heat waves such as those occurring in 1988. He never said that the particular drought of 1988 could have been caused by the greenhouse effect, but only that such events would likely increase as the planet heated up--not a very controversial view, and one that has been voiced by many scientists over the past 10 years.[3]

---

*Edited excerpts from testimony given before the Subcommittee on Oceanography and the Great Lakes of the Committee on Merchant Marine and Fisheries, U.S. House of Representatives, May 4, 1989. Any opinions, findings, conclusions, or recommendations expressed in this testimony are those of the author and do not necessarily reflect the views of the National Science Foundation.

ASPECTS OF THE DEBATE VISIBLE TO THE PUBLIC

Unfortunately, the widespread publicity following the drought in general and that hearing in particular led to the erroneous impression that Hansen and many other climatologists believed that the greenhouse effect was responsible for the drought conditions in 1988. Neither Hansen nor any other qualified atmospheric scientist I am aware of has ever made such a statement. In a recent letter to the New York Times Hansen commented, "as I testified to the Senate during the 1988 heat wave, the greenhouse effect cannot be blamed for a specific drought, but it alters the probabilities. Our climate model, tested by simulations of climate on other planets and past climates on Earth, indicates that the greenhouse effect is now becoming large enough to compete with natural climate variability."[4] Hansen did, however, suggest that the warming trend in the 1980s was very likely to have been caused by the greenhouse effect. Therefore, his "99 percent" statement was often confused with the idea that the drought was directly caused by global warming. The latter is obviously absurd, given that the amount of greenhouse gas increase from 1987 to 1988 is very small relative to the major climatic difference between those 2 years.

The strongest statement one can make responsibly, given the uncertainties, is that increasing the global average temperature could increase evaporation stress, thereby slightly altering the odds toward increased drought. However, as is well known by all atmospheric scientists, droughts are generally a result of unusual patterns of atmospheric circulation, whose causes are not clear cut and most often are ascribed to unusual temperature patterns in the oceans.

Later in 1988 that hypothesis of cause and effect was indeed reaffirmed by calculations made by Kevin Trenberth and his colleagues at the National Center for Atmospheric Research. Unfortunately, many atmospheric scientists were unaware of the exact statements made by Hansen and others who had testified at various hearings during the summer of 1988, and therefore they accepted the widespread perception, reported repeatedly in the media, that such witnesses believed the greenhouse effect and the drought were cause and effect.

This misperception caused an angry and understandable counterreaction. For example, the U.S. National Climate Program Office helped to sponsor a workshop on climate trends at the National Academy of Sciences on September 29, 1988. It focused on several difficulties with the observational record used to reconstruct global temperature trends, including the fact that many thermometers have had cities grow around them, which causes an unnatural urban heat island effect. Other errors were noted, such as stations that had moved out to airports or up or down mountains. I will discuss the seriousness of these issues below.

While this workshop was taking place, the calculations by Trenberth and his colleagues were announced. These suggested that unusually cool surface temperatures in the equatorial Pacific (actually, the cool water was flanked by warm water) may have been responsible for distorting the jet stream, thereby steering storms up and out of the United States in the spring of 1988 and contributing to the intense drought in the summer. This result was cited frequently in headline after headline that stated,

in effect, "Drought of '88 not caused by greenhouse effect after all."
The widespread association that was created from the summer's coverage
thereby led to what was, in essence, an artificial debate. In an article
published in Science, Trenberth and his colleagues quite rightly argued
that "climate simulations indicate that a doubling of carbon dioxide
concentrations could increase the frequency of summer droughts over North
America. Thus, the greenhouse effect may tilt the balance such that
conditions for droughts and heat waves are more likely, but it cannot be
blamed for an individual drought."[5] That statement is clearly reasonable and quite consistent with those of most other scientists who speak
on this issue.

The trial by media of the greenhouse effect was thus a nonscientific
issue from the very beginning. Nevertheless, many newspaper pieces continue to appear from scientists and others criticizing the "hysteria"
being generated by some (usually unnamed) scientists speaking in public
forums on this issue. See, for example, S. Fred Singer's "Fact and Fancy
on the Greenhouse Effect" (Wall Street Journal, August 30, 1988) and
Woods Hole Oceanographic Institution statistician Andrew Solow's "The
Greenhouse Effect: Hot Air in Lieu of Evidence" (International Herald
Tribune, December 29, 1988). But the most visible recent such article,
I believe, was University of Virginia climatologist Patrick Michaels'
article "The Greenhouse Climate of Fear" (Washington Post, January 8,
1989). Michaels listed a number of issues he believed were improperly
reported or underreported in the press, giving examples of "a few recent
revelations that somehow got lost with the ozone." He disparaged
Hansen's June testimony forecasting that 1988 would be the warmest year
on record, even though recent evidence now suggests that indeed it was.[6]
He went on to refute the greenhouse explanation of drought by citing
Trenberth's work to the effect that the drought of 1988 was caused by
cold tropical ocean temperatures, something that was never doubted by
other scientists. He cited the very recent results of National Oceanic
and Atmospheric Administration scientist Tom Karl, who, Michaels said,
"arguably knows more about regional climate variation than anyone in the
world." Michaels described Karl as saying that the NASA-GISS calculations for warming over the United States were too high by nearly 1°F for
this century because of the urban heat island effect. Because of this,
Michaels continued, "there may have been no [his emphasis] global
warming to speak of during the last century. Karl's findings surprised
none of us who daily toil with the data. But it should be a major shock
to those who are using those figures for policy purposes. Is it irre-
sponsible to point this out in public?"

The results of the Goddard Institute for Space Studies and Climatic
Research Unit (CRU) are reproduced here as Figure 22.1. The GISS record
shows a warming of about 0.8°C in the past 100 years and the CRU record a
global warming of about 0.6°C over the same period. Both include some of
the same thermometers, but the CRU record adds an oceanic data set not
included in the GISS record. Both groups were aware of and had attempted
to account for the artificial heating in urban areas due to the growth of
cities around such thermometers during the past century. The questions,
therefore, are how well they were able to make corrections for this urban
effect and whether that effect does indeed, as Michaels implied,

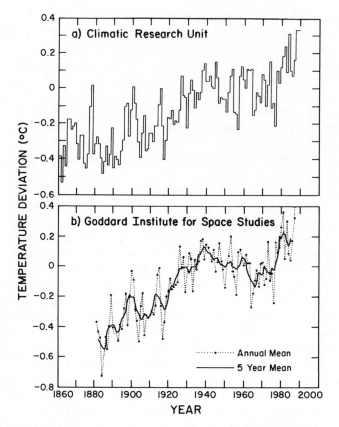

FIGURE 22.1 A comparison of the global surface temperature trends of the
past 100 years constructed (a) from land and island stations and ocean
surface temperature data sets at the Climatic Research Unit and (b) from
a similar set of stations (minus the ocean surface temperature data set)
at the Goddard Institute for Space Studies. (Sources: P. D. Jones and
T. M. L. Wigley, personal communication, 1988; J. Hansen and S. Lebedeff. 1988. Geophys. Res. Lett. 15:323.)

invalidate the notion that the globe has been warming for the past
century.

     Neither of these groups had easy access to the vast network of rural
stations that Karl had at the National Climate Center in Asheville, North
Carolina, and that enabled him to check the accuracy of predictions of
U.S. temperature trends made by both the GISS and CRU groups. What
Tom Karl and Phil Jones found was that the GISS results had overestimated
the warming trend in the United States by about 0.3°C (an error more than
30 percent lower than that implied by Michaels), but the East Anglia
(England) group had only overestimated the warming trend by about
0.15°C.[7] Recall that the GISS global record suggested a warming of
about 0.8°C in the past 100 years, and the East Anglia group record a
warming of about 0.6°C or so.

Therefore, if one makes the very conservative assumption that the entire world record suffers as strong an urban heat island bias as the United States, and thus applies Karl's corrections to the global record (even though they only apply to the United States), then that correction procedure simply reduces Hansen's record and the CRU record to about 0.5°C (0.9°F) global warming for the past century. This is not consistent with the frequent statements in the press that urban biases are likely to eliminate the global warming trend over the past century. Moreover, this is not believed by Karl or any other knowledgeable atmospheric scientists I am aware of. Indeed, Solow has not contested the existence of a global warming trend, only its relevance to the increased greenhouse gas burden of the past 100 years. And in an interview with Science magazine's Richard Kerr, Karl commented that "the long-term global warming is something on the order of 0.4°C during the past century. Is the [urban] bias 0.05°C or 0.2°C? The chances that it is the same size as the [global] warming are pretty remote. It is a matter of adjusting the rate of rise, not questioning the rise itself" (Science 243, 603, 1989).

Another example of an overblown debate with little relevance to global warming occurred in mid-January 1989, when a number of Department of Commerce scientists (including Karl) published a U.S. temperature trend record that got widespread press coverage (Weekly Climate Bulletin, No. 89/02, Jan. 14, 1989). They pointed out that there had been no net warming in the United States over the twentieth century. This, too, was greeted with many headlines that the greenhouse effect had not happened in the United States. Good articles (albeit with questionable headlines) appeared on this subject,[8] quoting a number of sources, including the authors of the North American study, to the effect that the United States occupies only a small percentage of the world's area, and that there are almost continent-sized regions that have been warming substantially, as well as continent-sized regions that have been cooling, over the past few decades. Therefore, it is just as foolish to draw global inferences from looking at the small fraction of the earth represented by the United States as it is to predict the outcome of a national election by looking at trends in only one or two states. Nevertheless, an artificial debate has grown over the validity of the greenhouse effect as a result of this publication.

SOME ELEMENTS OF THE SCIENTIFIC DEBATE

What then, has the debate visible to the public to do with the actual scientific debate? I am afraid the answer is often "too little." Virtually all atmospheric scientists believe that the greenhouse effect as a scientific proposition is well established and essentially beyond question. The trapping of heat near the earth's surface from the presence of gases such as water vapor, carbon dioxide, methane, and chlorofluorocarbons has been established from literally millions of measurements over the past century. It is also beyond debate that carbon dioxide has increased about 25 percent since the Industrial Revolution, and methane substantially more than that. It is also virtually certain that the CO_2 increases are attributable to human activities, principally the burning of fossil fuels and deforestation, and that methane increases are initially connected with agriculture and land use.

There is controversy, but it involves how much trace gases will increase over the next century. Given typical assumptions of growth in fossil fuel use, population, and standards of living into the future, the uncertainties in the greenhouse gas projections are roughly on the order of a factor of 3. Uncertainty in the global climate response to a given increase in trace gases is also typically on the order of a factor of 2 to 3. Taken together, these uncertainties explain why most national and international assessments suggest that a warming ranging from as little as 1°C to as much as 10°C (1.8 to 18°F) is possible by the end of the next century. A few believe that the warming could be even greater or that a cooling could be triggered, but these views represent a very small segment of scientists. Major assessments put the most probable global temperature increase over the next 50 years or so at several degrees Celsius.[9]

Less certain than projections of global average temperature increase is the regional distribution of climate changes. Such changes could have major impacts on water resources, agriculture, forests, urban infrastructure, human health, navigation, or coastal planning.[10] Whether predictions of drying in the midcontinental United States and loss of comparative advantage for U.S. agriculture, which are frequently inferred from climatic models, will be confirmed is simply not known now. Considering the current state of the art, it is doubtful that reliable forecasts of regional climatic changes will be available for another 20 years. That distant date might be accelerated substantially if both present and expanded research efforts were carefully organized, but even then, we cannot guarantee scientific consensus on the reliability of regional predictions of climate change, even in the time frame of a decade. But I believe it is worth a try to make the effort to improve the reliability of regional climate predictions, for such information would help to put evolving policymaking on a firmer factual basis.

VALUE JUDGMENTS AND POLICYMAKING

Finally, the greatest lack of scientific consensus occurs over whether present uncertainties justify an immediate policy response--which is a value judgment, not an issue resolvable by scientific methods. The policy process is advanced when scientists provide what they are technically competent to offer: estimates of specific consequences of greenhouse gas buildups and their likelihood of occurrence--even if estimates of the latter are based on intuitive judgments of technical experts. Any statements beyond that are the personal opinions of those scientists. Although I believe that scientists, like all citizens, are entitled to opinions on how to deal with those probabilities and consequences, we must always be scrupulous to point out that such opinions are personal value judgments.

Whether a certain amount of knowledge justifies urgent action or delay depends on whether one fears investing present resources as a hedge against future change more than one fears future change descending without some attempt having been made either to slow it or to make investments to adapt to it more effectively. In my value system, the latter is a clearly preferable course, for I believe in insuring against potential catastrophic loss at both the personal and national levels.

Waiting for more evidence before acting also represents a value judgment. But the odds of rapid, potentially unprecedented--even catastrophic--climate change are almost certainly in the first decimal place of probability, and I believe the likelihood of a several-degrees-Celsius warming in 50 years is a better-than-even bet. The cost of overreaction is a legitimate issue, but so too is the cost of under-reaction.

The more rapidly climate changes evolve, the more difficult it will be for societies to adapt; and it may be impossible for natural ecosystems to adapt without substantial dislocations or extinctions of some species. The more rapidly greenhouse gases build up, the more difficult it will be for scientists to forecast the outcome reliably. The less we know about the future, the less easily we will be able to adapt--or even to take advantage of some of the changes that will occur. If we choose to wait for that added degree of certainty before implementing preventive policies, the delay will not be cost-free, for it must occur at the price of forcing living things to adapt to much greater change than what might occur if we were to act today to slow the change or to invest affirmatively to make our future adaptations easier. Wrangling over a few tenths of a degree in the historic global temperature trends in this hemisphere or that will not change the fact that postponing action is a basic gamble with our environmental future. Quite simply, the "bottom line" of the evolving greenhouse gas buildup is that we are insulting the environment at a rate greater than our ability to predict the consequences and that, under these conditions, surprises are virtually certain.

NEED FOR ONGOING SCIENTIFIC INVESTIGATION

Continued Observation and Monitoring

To establish that the greenhouse effect signal has clearly been detected in the climatic record, we will require another decade or possibly 2 to be sure that the warming of the 1980s (which does appear to be the warmest decade recorded on a global basis) will in fact continue into the 1990s and beyond. The nature of regional climate fluctuations and global trends during the past century is very difficult to establish to a high degree of accuracy because of (1) missing or faulty temperature data as well as (2) incomplete data on other possible competitive causes of climate change, such as changes in the energy output from the sun, human dust, or other pollutants.[11]

Model Validation

The strongest argument for concern about the rapid buildup of greenhouse gases comes not from the small global climatic trends and large regional climatic noise of the recent past, but rather from the well-validated nature of the physical processes that have long been understood to accompany increases in infrared-radiation-trapping (i.e., greenhouse) gases.

Let us consider in more detail the important issue of model valida- tion. Perhaps the most perplexing question about climate models is whether they can ever be trusted enough to provide grounds for altering social policies, such as those governing carbon dioxide emissions. How can models so fraught with uncertainties be verified? There are actually several methods. None of them is sufficient on its own, but together they can provide significant (albeit largely circumstantial) evidence of a model's credibility.

The first method is to check the model's ability to simulate today's climate. The seasonal cycle is one good test because the temperature changes involved are large--several times larger, on the average, than the change from an ice age to an interglacial period. General circula- tion models (GCMs) do remarkably well at mapping the seasonal cycle, which strongly suggests they are on the right track. The seasonal test is encouraging as a validation of "fast physics," such as changes in cloudiness. However, it does not indicate how well a model simulates slow processes, such as changes in deep ocean circulation, that may have important long-term effects.

A second method of verification is to isolate individual physical components of a model, such as its parameterizations, and test them against either a high-resolution submodel or real data from the field. For example, one can check whether a model's parameterized cloudiness matches the level of cloudiness appropriate to a particular grid box. Or one can test a GCM's grid cloudiness against an isolated mesoscale model. The problem with the former test is that it cannot guarantee that the complex interactions of many individual model components are properly treated. The GCM may be good at predicting average cloudiness but bad at representing cloud feedback. In that case the simulation of the overall climatic response to, say, increased carbon dioxide is likely to be in- accurate.

A third method for determining overall, long-term simulation skill is to check a model's ability to reproduce the diverse climates of the ancient earth or even of other planets. Paleoclimatic simulations of the Mesozoic Era, glacial/interglacial cycles, or other extreme past climates help in understanding the coevolution of the earth's climate with living things. As verifications of climate models, however, such simulations are also crucial to estimating both the climatic and biological future.[12]

Overall validation of climatic models thus depends on constant ap- praisal and reappraisal of performance in the above categories. Also important are a model's response to such century-long forcings as the 25 percent increase in carbon dioxide and other trace greenhouse gases since the Industrial Revolution. Indeed, most climatic models are sensitive enough to predict that warming of at least 1°C should have occurred during the past century. The precise "forecast" of the past 100 years also depends on how a model accounts for such factors as changes in the solar constant, sulfate aerosols, or volcanic dust.[13]

Indeed, as recent data show, the typical prediction of a 1°C warming is broadly consistent but somewhat larger than the amount of warming actually observed. Possible explanations for the discrepancy include the following:[14] (1) state-of-the-art models are too sensitive to increases in trace greenhouse gases by a rough factor of 2; (2) modelers have not properly accounted for such competitive external forcings as volcanic dust or changes in solar energy output--which could have caused fluctuations of up to several tenths of a degree Celsius; (3) modelers have not accounted for other external forcings such as regional tropospheric aerosols from agricultural, biological, and industrial activity;[15] (4) modelers have not properly accounted for internal processes that could lead to stochastic or chaotic behavior; (5) modelers have not properly accounted for the large heat capacity of the oceans taking up some of the heating of the greenhouse effect and delaying, but not ultimately reducing, warming of the lower atmosphere; (6) both present model forecasts and observed climatic trends could in fact be consistent because models are typically run for equivalent doubling of carbon dioxide, whereas the world has only experienced one-quarter of this increase, so that nonlinear processes have been properly modeled and have produced a sensitivity appropriate for a doubling but not for a 25 percent increase; and (7) the incomplete and inhomogeneous network of thermometers has underestimated actual global warming during this century.

Despite these explanations, the empirical test of model predictions against a century of observations certainly is consistent to a rough factor of 2. This test is reinforced by the good simulation by most climatic models of the seasonal cycle, diverse ancient paleoclimates, hot conditions on Venus, cold conditions on Mars (both well simulated), and the present distribution of climates on earth. When taken together, these verifications provide a strong circumstantial case that the modeling of sensitivity of the global surface air temperature to greenhouse gases is probably valid within roughly 2-fold. Another decade or 2 of observations of trends in the earth's climate should produce signal-to-noise ratios sufficiently obvious that most scientists will know whether present estimates of climatic sensitivity to increasing trace gases have been accurate or not.

CONCLUDING REMARKS

It is my personal belief that we clearly know more than enough to actively pursue actions that both slow the rate of buildup of greenhouse gases and at the same time help solve other societal problems--what has been called the "tie-in strategy." In particular, individuals, firms, and nations should pursue those activities that make general good sense, regardless of whether the greenhouse effect turns out to be much less serious than currently contemplated.[16] For example, stepped-up investments in energy production and end-use efficiency, accelerated testing of nonfossil fuel alternatives, development of more widely climate-adapted crop strains, added flexibility in the management of water systems, and coastal planning to deal with rising sea levels and storm surges would all contribute to solving existing societal problems even if further climate change does not occur. Therefore it seems prudent to consider implementing first those policies that have multiple benefits, including that of buying insurance against the real possibility of large and potentially catastrophic climate change.

I hope that recent shrill or irrelevant debates or headlines do not mask the already large scientific consensus that exists over the basic physical phenomenon known as the greenhouse effect, a scientific proposition over which I have heard virtually no scientific dissent.

In any case, regardless of whether society chooses vigorous prevention policies as a response to global warming, nearly all analysts agree that some growth in greenhouse gases will continue into the twenty-first century. Therefore a vigorous program of interdisciplinary research on earth systems, what has been called "Global Change,"[17] will help society adapt more effectively to the global changes that appear inevitable.

NOTES

1. For example, S.H. Schneider and R. Londer, 1984, The Coevolution of Climate and Life (Sierra Club Books, San Francisco), Chapter 8, pp. 294-365; or S.H. Schneider, 1989, Global warming: Scientific reality or political hype? Pp. 53-57 in Global Warming. Hearings before the Subcommittee on Energy and Power of the Committee on Energy and Commerce, U.S. House of Representatives, February 21, 1989 (U.S. Government Printing Office, Washington, D.C.); S.H. Schneider, 1989, Global Warming: Are We Entering the Greenhouse Century? (Sierra Club Books, San Francisco, 317 pp.).

2. J.E. Hansen, 1988, prepared statement, The Greenhouse Effect: Impacts on Current Global Temperature and Regional Heat Waves. In Greenhouse Effect and Global Climate Change, hearing before the Committee on Energy and Natural Resources, U.S. Senate, Washington, D.C., June 23, pp. 42-79.

3. For example, L.O. Mearns, R.W. Katz, and S.H. Schneider, 1984, Changes in the probabilities of extreme high temperature events with changes in global mean temperature, J. Clim. and Appl. Meteorol. 23, 1601-1613, pointed out that if temperatures were to increase by only 3°F, and nothing else in the climate system changed, then the probability of a heat wave in which 5 or more days in a row in July saw afternoon temperatures greater than 95°F would increase in Washington, D.C., from a present probability of around 1 in 6 to a future probability of around 1 in 2. In Des Moines, Iowa, the respective probabilities would change from 1 in 18 to 1 in 5. These changes in the odds on "climatic dice" are what would be expected if only temperature increased by 3°F.

4. J. Hansen, letter to the New York Times, January 11, 1989.

5. K.E. Trenberth, G.W. Branstator, and P.A. Arkin, 1988, Origins of the 1988 North American drought, Science 242:1640-1645.

6. For example, in a page 1 story, Philip Shabecoff reports in the February 4, 1989, New York Times that the Climatic Research Unit in East Anglia, England, has concluded that 1988 did, indeed, set a record as the year with the warmest global temperature. P.D. Jones of that unit has confirmed by correspondence the accuracy of this story.

7. T.R. Karl and P.D. Jones, 1989, Urban bias in area-averaged surface air temperature trends, Bulletin of the American Meteorological Society 70:265-270.

8. P. Shabecoff, U.S. Data Since 1895 Fail to Show Warming Trend, New York Times, January 26, 1989, p. 1.

9. See S.H. Schneider, 1989, The greenhouse effect: science and policy, Science 243:771-781, for references to many such assessments.

10. J.B. Smith and D. Tirpak, eds., October 1988, The Potential Effects of a Global Climate Change on the United States: Draft Report to Congress, Vols. I and II (Environmental Protection Agency, Washington, D.C.).

11. S.H. Schneider, 1989, Science 243:775-776.

12. S.H. Schneider and R. Londer, 1984: The Coevolution of Climate and Life (Sierra Club Books, San Francisco).

13. S.H. Schneider and C. Mass, 1975, Volcanic dust, sunspots, and temperature trends, Science 190:741-746; R.L. Gilliland and S.H. Schneider, 1984, Volcanic, CO_2, and solar forcing of northern and southern hemisphere surface air temperatures, Nature 310:38-41; J. Hansen, D. Johnson, A. Lacis, S. Lebedeff, P. Lee, D. Rind, and G. Russell, 1981, Climate impact of increasing atmospheric carbon dioxide, Science 213:957-966.

14. S.H. Schneider, 1989, Science 243:775-776.

15. Wigley, T.M.L., 1989, Possible climate change due to SO_2-derived cloud condensation nuclei, Nature 339:365-367.

16. See chapter 8 of S.H. Schneider, 1989, Global Warming: Are We Entering the Greenhouse Century? (Sierra Club Books, San Francisco, 317 pp.).

17. National Research Council, 1988, Toward an Understanding of Global Change: Initial Priorities for U.S. Contributions to the International Geosphere-Biosphere Program (National Academy Press, Washington, D.C.).

APPENDIXES

PROGRAM--FORUM ON GLOBAL CHANGE AND OUR COMMON FUTURE

Tuesday, May 2, 1989

8:00 a.m.   Registration

9:00 a.m.   Welcome
             Thomas Malone, St. Joseph College

9:10 a.m.   Society's Stake in Global Change
             William Ruckelshaus, Browning-Ferris Industries

Understanding Global Change: The Science
Chair: Thomas Malone, St. Joseph College

9:45 a.m.   Historical Perspectives: Climate Changes
             Throughout the Millennia
             John Kutzbach, University of Wisconsin

10:15 a.m.   Break

10:25 a.m.   Understanding Global Change: Earth as a System
             Francis Bretherton, University of Wisconsin

11:00 a.m.   <u>Panel: The Earth System</u>
             Moderator: Digby McLaren, Royal Society of Canada

             Atmosphere
             Michael McElroy, Harvard University

             Oceans
             James McCarthy, Harvard University

             Terrestrial Ecosystems
             Peter Vitousek, Stanford University

             Human Dimensions
             Roberta Miller, National Science Foundation

223

12:45 p.m.	Lunch
2:15 p.m.	Human Causes of Global Change B.L. Turner II, Clark University
2:45 p.m.	Panel: Consequences Moderator: Robert McC. Adams, Smithsonian Institution

Greenhouse Warming
Jerry Mahlman, Geophysical Fluid Dynamics
Laboratory/National Oceanic and Atmospheric
Administration

Stratospheric Ozone Depletion
Susan Solomon, Environmental Research
Laboratory/National Oceanic and Atmospheric
Administration

Break

Deforestation
Eneas Salati, Escola Superior de Agricultura,
Sao Paulo, Brazil

Acid Deposition
David Schindler, Department of Fisheries and Oceans, Canada

Implications for Life on Earth
Paul Ehrlich, Stanford University

5:30 p.m.	Recess
7:30 p.m.	Keynote Address and Franklin Lecture Global Change and Our Common Future Mme. Gro Harlem Brundtland, Prime Minister of Norway

Wednesday, May 3, 1989

Impacts of Global Change
Chair: Robert Hoffmann, Smithsonian Institution

9:00 a.m.	What Does Global Change Mean for Society? Lester Brown, Worldwatch Institute
9:30 a.m.	Panel: Impacts Moderator: Jose Goldemberg, University of Sao Paulo, Brazil

Agriculture and Water Resources
Theodore Hullar, University of California, Davis

Break

        Biodiversity
        Robert Peters, World Wildlife Fund

        Sea Level
        James Broadus, Woods Hole Oceanographic Institution

        Industry
        Hugh Wynne-Edwards, Alcan, Canada

11:30 a.m.   Implications of Global Change for Science Policy
              Robert Corell, National Science Foundation

12:00 noon   Lunch

Implications for Public Policy
Chair: Thomas Lovejoy, Smithsonian Institution

 1:30 p.m.   Options for Action
              Martin Holdgate, International Union for Conservation of
              Nature and Natural Resources

 2:00 p.m.   View from the North
              Digby McLaren, Royal Society of Canada

 2:30 p.m.   View from the South
              Marc Dourojeanni, The World Bank

 3:00 p.m.   Break

 3:10 p.m.   Panel: Public Policy
              Moderator: Jessica Mathews, World Resources Institute
              The Honorable Charles Caccia, member of Parliament, Canada
              The Honorable John Chafee, U.S. Senate
              William A. Nitze, U.S. Department of State
              Paulo Nogueira-Neto, University of Sao Paulo, Brazil
              The Honorable Timothy Wirth, U.S. Senate

 5:30 p.m.   Recess

Summary Panel

 7:30 p.m.   Moderator: Thomas Malone, St. Joseph College
              Alan Hecht, National Oceanic and Atmospheric Administration
              Rafael Herrera, Instituto Venezolano de Investigaciones
                  Cientificas, Venezuela
              John Holdren, University of California, Berkeley
              Thomas Lovejoy, Smithsonian Institution
              Stephen Schneider, National Center for Atmospheric Research
              Anne Whyte, International Development Research Centre, Canada

APPENDIX B

COMMITTEE ON GLOBAL CHANGE AND OVERSIGHT COMMITTEE MEMBERS

COMMITTEE ON GLOBAL CHANGE
(U.S. National Committee for the IGBP)

Harold Mooney, Stanford University, <u>Chairman</u>
Paul G. Risser, University of New Mexico, <u>Vice Chairman</u>
D. James Baker, Joint Oceanographic Institutions, Inc.
Francis P. Bretherton, University of Wisconsin
Kevin C. Burke, National Research Council
William C. Clark, Harvard University
Margaret B. Davis, University of Minnesota
Robert E. Dickinson, National Center for Atmospheric Research
John Imbrie, Brown University
Robert W. Kates, Brown University
Thomas F. Malone, St. Joseph College
Michael B. McElroy, Harvard University
Berrien Moore III, University of New Hampshire
Ellen S. Mosely-Thompson, Ohio State University
Piers J. Sellers, University of Maryland

<u>Ex-Officio Members</u>
<u>U.S. Members, ICSU Special Committee for the IGBP</u>
John A. Eddy, University Corporation for Atmospheric Research
James J. McCarthy, Harvard University
S. Ichtiaque Rasool, National Aeronautics and Space Administration

<u>Staff</u>
John S. Perry, <u>Staff Director</u>
Ruth S. DeFries, <u>Staff Officer</u>

227

AD HOC COMMITTEE FOR OVERSIGHT OF
THE COMMITTEE ON GLOBAL CHANGE

Norman Hackerman, Robert A. Welch Foundation (Chairman, Commission on Physical Sciences, Mathematics, and Resources), Convenor
Robert McC. Adams, Smithsonian Institution (Chairman, Commission on Behavioral and Social Sciences and Education)
Bruce Alberts, University of California, San Francisco (Chairman, Commission on Life Sciences)
William Gordon, Rice University (Foreign Secretary, National Academy of Sciences)

Staff
John Burris, Executive Director, Commission on Life Sciences
Victor Rabinowitch, Executive Director, Office of International Affairs
Myron Uman, Acting Executive Director, Commission on Physical Sciences, Mathematics, and Resources
Suzanne Woolsey, Executive Director, Commission on Behavioral and Social Sciences and Education